Hyperparameter Optimization in Machine Learning

Make Your Machine Learning and Deep Learning Models More Efficient

Tanay Agrawal

Apress®

Hyperparameter Optimization in Machine Learning

Tanay Agrawal
Bangalore, Karnataka, India

ISBN-13 (pbk): 978-1-4842-6578-9 ISBN-13 (electronic): 978-1-4842-6579-6
https://doi.org/10.1007/978-1-4842-6579-6

Managing Director, Apress Media LLC: Welmoed Spahr
Acquisitions Editor: Celestin Suresh John
Development Editor: Matthew Moodie
Coordinating Editor: Aditee Mirashi

Cover designed by eStudioCalamar

Cover image designed by Freepik (www.freepik.com)

Distributed to the book trade worldwide by Springer Science+Business Media New York, 1 New York Plaza, Suite 4600, New York, NY 10004-1562, USA. Phone 1-800-SPRINGER, fax (201) 348-4505, e-mail orders-ny@springer-sbm.com, or visit https://www.springeronline.com. Apress Media, LLC is a California LLC and the sole member (owner) is Springer Science + Business Media Finance Inc (SSBM Finance Inc). SSBM Finance Inc is a **Delaware** corporation.

For information on translations, please e-mail booktranslations@springernature.com; for reprint, paperback, or audio rights, please e-mail bookpermissions@springernature.com.

Apress titles may be purchased in bulk for academic, corporate, or promotional use. eBook versions and licenses are also available for most titles. For more information, reference our Print and eBook Bulk Sales web page at https://www.apress.com/bulk-sales.

Any source code or other supplementary material referenced by the author in this book is available to readers on GitHub via the book's product page, located at https://www.apress.com/978-1-4842-6578-9. For more detailed information, please visit https://www.apress.com/source-code.

Printed on acid-free paper

*This book is dedicated to my parents
and my grandparents.*

Table of Contents

About the Author

Tanay Agrawal is a deep learning engineer and researcher who graduated in 2019 with a bachelor of technology from SMVDU, J&K. He is currently working at Curl Hg on SARA, an OCR platform. He is also advisor to Witooth Dental Services and Technologies. He started his career at MateLabs working on an AutoML Platform, Mateverse. He has worked extensively on hyperparameter optimization. He has also delivered talks on hyperparameter optimization at conferences including PyData, Delhi and PyCon, India.

About the Technical Reviewer

Manohar Swamynathan is a data science practitioner and an avid programmer, with over 14 years of experience in various data science–related areas that include data warehousing, business intelligence (BI), analytical tool development, ad hoc analysis, predictive modeling, data science product development, consulting, formulating strategy, and executing analytics program. He's had a career covering the life cycle of data across different domains such as US mortgage banking, retail/e-commerce, insurance, and industrial IoT. He has a bachelor's degree with a specialization in physics, mathematics, and computers and a master's degree in project management. He's currently living in Bengaluru, the Silicon Valley of India.

Acknowledgments

I would like to thank Kailash Ahirwar (CTO, MateLabs) for being a great mentor. A big thanks to teams from both MateLabs and Curl HG for their constant support. I am grateful to Akruti Acharya and Jakub Czakon for their insightful inputs while writing this book. I would also like to thank Paankhi Agrawal, Sahil Sharma, Dr. H.-Y. Amanieu, Anubhav Kesari, Abhishek Kumar, Amog Chandrashekar, and others. This book wouldn't have been possible without all the love and encouragement from my family.

Foreword 1

I have to admit that tweaking parameters by hand was something that I really enjoyed when I trained my first ML models. I would change a parameter, run my training script, and wait to see if the evaluation score improved. One of those guilty pleasures.

But as I spent more time in the ML world, I understood that there are other, more impactful areas where I could spend my time. I realized that I could (and should) outsource parameter tweaking somewhere.

I learned about random search and started using it in my projects. At some point, I felt I could do better than random search and started reading about more advanced hyperparameter optimization algorithms and libraries.

A lot of articles I found where pretty shallow and basic, but I remember reading this deep, hands-on yet easy-to-follow article about Hyperopt, one of the most popular HPO libraries. It was written by Tanay Agrawal. That article probably still is one of the more valuable articles I've ever read on the subject. I mentioned it in one of my blog posts and this is how we met.

When Tanay told me that he was writing a book about hyperparameter optimization, without hesitation, I proposed to review it. I am not going to lie, I really wanted to read it before anyone else! To my surprise, Tanay agreed and I was able to give a few notes here and there.

This book truly is a deep dive into the theory and practice of hyperparameter optimization. I really like how it explains theory deeply but not in an overly complex way. The practical examples are centered on the libraries and frameworks that are heavily used today, which makes this book current and, most importantly, useful.

I recommend this book to any ML practitioner who wants to go beyond the basics and learn the why, how, and what of hyperparameter optimization.

Jakub Czakon
Senior Data Scientist
Neptune.ai

Foreword 2

In this book, Tanay takes you on an interactive journey—in the most literal sense, as each line of code can be run in a notebook—of the depths of hyperparameters. It helps anyone to quickly get started on tuning and improving their deep learning project with any library they choose to use.

The author mindfully covers the inner workings of hyperparameters in ML models in a thorough but accessible fashion, which will allow you to understand and build upon them using different libraries. The book also demystifies the blackest of the black box: hyperparameter optimization in automated machine learning.

It's a friendly guide to a complicated subject, and yet it's full of cutting-edge gems that even advanced practitioners will love.

Akruti Acharya
Data Scientist
Curl HG

Introduction

Choosing the right hyperparameters when building a machine learning model is one of the biggest problems faced by data science practitioners. This book is a guide to hyperparameter optimization (HPO). It starts from the very basic definition of *hyperparameter* and takes you all the way to building your own AutoML script using advance HPO techniques. This book is intended for both students and data science professionals.

The book consists of five chapters. Chapter 1 helps you to build an understanding of how hyperparameters affect the overall process of model building. It teaches the importance of HPO. Chapter 2 introduces basic and easy-to-implement HPO methods. Chapter 3 takes you through various techniques to tackle time and memory constraints. Chapters 4 and 5 discuss Bayesian optimization, related libraries, and AutoML.

The intent of this book is for readers to gain an understanding of the HPO concepts in an intuitive as well as practical manner, with code implementation provided for each section. I hope you enjoy it.

CHAPTER 1

Introduction to Hyperparameters

Artificial intelligence (AI) is suddenly everywhere, transforming everything from business analytics, the healthcare sector, and the automobile industry to various platforms that you may enjoy in your day-to-day life, such as social media, gaming, and the wide spectrum of the entertainment industry. Planning to watch a movie on a video-streaming app but can't decide which movie to watch? With the assistance of AI, you might end up watching one of the recommendations that are based on your past movie selections.

Machine learning is a subset of AI that involves algorithms learning from previous experiences. In some cases, machine learning has achieved human-level accuracy. For example, state-of-the-art deep neural networks (DNNs) perform as well as humans in certain tasks, such as image classification, object detection, and so forth, although this is not the same as simulating human intelligence (but it's a step).

In machine learning algorithms, tuning hyperparameters is one of the important aspects in building efficient models. In this chapter you'll discover the meaning of the term *hyperparameter* and learn how hyperparameters affect the overall process of building machine learning models.

T. Agrawal, *Hyperparameter Optimization in Machine Learning*,
https://doi.org/10.1007/978-1-4842-6579-6_1

Introduction to Machine Learning

Machine learning is the study of algorithms which perform a task without explicitly defining the code to perform it, instead using data to learn. Machine learning enables algorithms to learn on their own without human intervention.

Tom M. Mitchell, a computer scientist and, at the time of writing, a professor at Carnegie Mellon University, defines machine learning as follows: "A computer program is said to learn from experience *E* with respect to some class of tasks *T* and performance measure *P* if its performance at tasks in *T*, as measured by *P*, improves with experience *E*."

Machine learning algorithms have several subdivisions based on the type of problem that needs to be solved. Here I will introduce you to three main types:

- *Supervised machine learning algorithms*: Labeled data is provided, we build a model over it to predict such labels given variables. As an example, suppose you want to purchase a spaceship. Several factors would help you to decide which spaceship to buy: cost, size of spaceship, build quality, whether it has hyperdrive, its weaponry system, and so on. Now we have data of hundreds of spaceships with such feature information and their price, so we build a model and predict the price. This comes under the regression problem. *Regression problems* have continuous target values, and if the target values are discrete, we call them *classification problems*. A third type of problem features time-stamps, *time series forecasting*, where the next data point is somewhat dependent on the previous information, so your algorithm needs to keep in memory information from the previous data points.

The image on the left in Figure 1-1-1 is an example of a classification problem. We need labeled data in order to learn to draw a seperation between them.

- *Unsupervised machine learning algorithms*: These kinds of problems do not have a target value. Suppose you have to group the hypothetical spaceships in clusters according to their features; you would use a *clustering algorithm* to do so. Unsupervised machine learning is used to detect patterns among the dataset. You don't know which cluster is which, but you do know that all the spaceships in one cluster are similar to each other; the right image in Figure 1-1-1 shows an example.

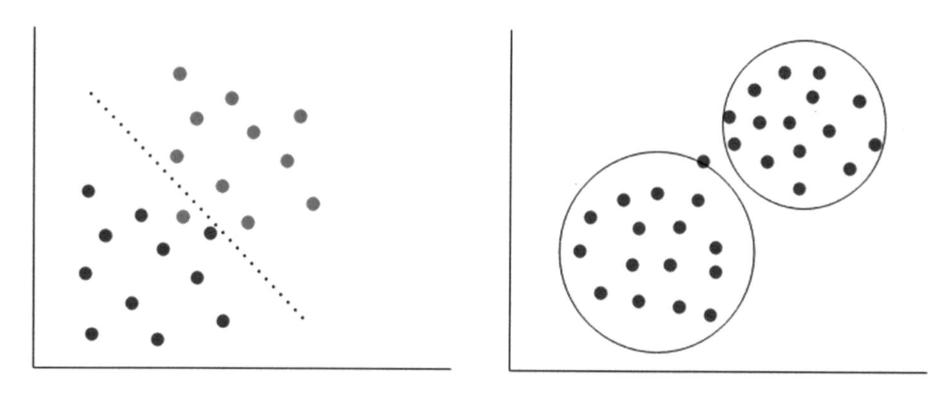

Figure 1-1-1. *Examples of supervised machine learning (left) and unsupervised machine learning (right)*

- *Reinforcement machine learning algorithms*: A reinforcement machine learning algorithm learns from the environment; if it performs well, it gets a reward, and the goal is to maximize the reward. For example, consider the Chrome "running dinosaur" game (go to `chrome://dino/` and press the spacebar). The dinosaur continuously runs toward obstacles. To increase your score, you have to press the spacebar at the precise time to make the dinosaur jump over the obstacles. Here those points are the reward and jumping is the variable that needs to be decided at the right time. In problems like this, we use reinforcement machine learning algorithms. The *Q-learning algorithm* is one example of a reinforcement machine learning algorithm. One of the most brilliant applications of reinforcement machine learning is a robot learning to walk through trial and error.

There's a lot more to machine learning. You need to be familiar with the basics of machine learning before jumping into hyperparameters and their optimization methodologies. If you are new to machine learning or if you want to brush up on the basic concepts, refer to Appendix I and Appendix II. Appendix I covers practical application of machine learning and some of its basic aspects. Appendix II gives you a brief introduction to fully connected neural networks and the PyTorch and Keras frameworks for implementation.

Understanding Hyperparameters

There are two kinds of variables when dealing with machine learning algorithms, depicted in Figure 1-2-1:

- *Parameters*: These are the parameters that the algorithm tunes according to dataset that is provided (you don't have a say in that tuning)
- *Hyperparameters*: These are the higher-level parameters that you set manually before starting the training, which are based on properties such as the characteristics of the data and the capacity of the algorithm to learn

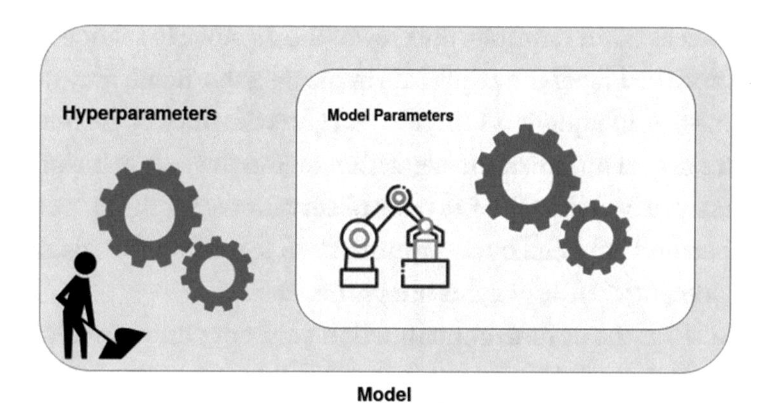

Figure 1-2-1. *The box inside represents model parameters, where the machine learning algorithm is at work. The outer box represents the hyperparameters, which we have to set before algorithm starts training*

I'll present a machine learning algorithm as an example to show you the difference between a parameter and a hyperparameter. Let's take a very basic algorithm, linear regression.

The hypothesis function in linear regression is as follows:

$$f(\Theta, \Theta_c) = \Theta . x + \Theta_c \qquad \text{(Equation 1.2.1)}$$

Here, x and Θ are vectors, x being a vector of features and Θ being the weights assigned to each feature, and Θ_c is a constant bias.

Let's consider as an example the classic problem of house price prediction. The price of a house is dependent on certain factors, including square footage of the house, number of bedrooms, number of washrooms, crime rate in the locality, distance from public transportation (bus station, airport, railway station), school district, distance to the nearest hospital, and so forth. All of these can be considered as features; that is, the x vector in our hypothesis function in Equation 1.2.1. The price of a house increases, for example, as the number of bedrooms increases and the square footage increases; these features would have positive weightage (Θ). The price of a house decreases, for example, the greater the distance to schools and hospitals and the higher the crime rate in the neighborhood; they would have negative Θ. In Equation 1.2.1, $f(\Theta, \Theta_c)$ gives the price of the house.

We can use an optimization algorithm to find the best value of Θ for each feature based on the previous observations. So, the Θ vector is controlled and adjusted by the optimization algorithm (for instance, gradient descent). These weights are *parameters*.

Let's discuss the optimization function gradient descent, which will help you to understand hyperparameters.

We'll start by assigning some random numbers (i.e., *weights*) to our parameters. For one observation, if we have vector x (with numerical values for each feature) and vector Θ (random numerical values for each weight), by using Equation 1.2.1, we get the value of $f(\Theta, \Theta_c)$. This will be our prediction, which will be some random value (p_1') because weights are random. And we have a true value of the house price (p_1).

We can calculate the difference, C_1 (for first observation), $|p_1\text{-}p_1'|$. This is a loss which we have to reduce. Similarly, if we calculate the average of summation of loss (C) for all the observation:

$$C(\Theta, \Theta c) = (1/n)\sum\left|p_i - p_i^{'}\right| \qquad \text{(Equation 1.2.2)}$$

Equation 1.2.2 is termed as loss function, the goal of optimization function is to reduce the value of C, so we can give more accurate predictions. The loss function is dependent on weights and bias, as depicted in Figure 1-2-2.

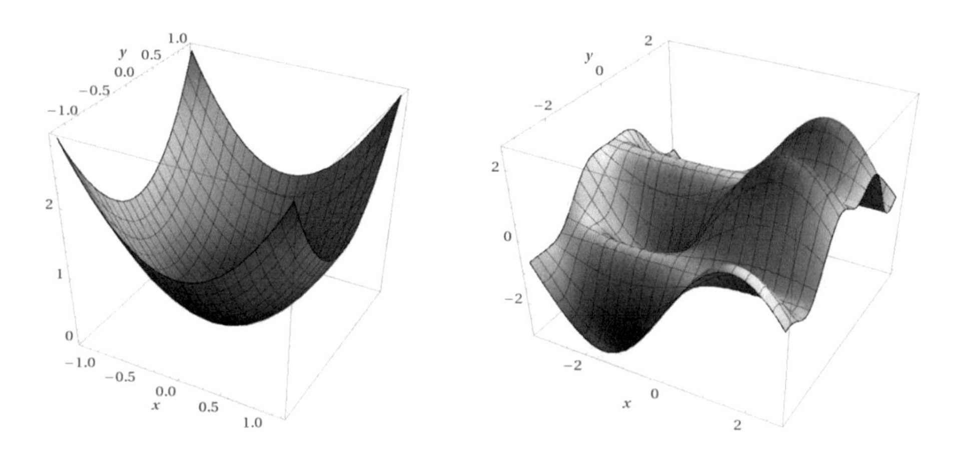

Figure 1-2-2. *Loss curves in three dimensions, with x and y axes being weights and z axis the loss*

Three-dimensional curves in Figure 1-2-2 can be possible representations of a loss function. Remember we started our weights and biases with random values; now we need to change those values such that loss moves to its minima. As per calculus, C changes as follows:

$$\Delta C \cong \Sigma(\delta C / \delta\Theta_i) * \Delta\Theta_i \qquad \text{(Equation 1.2.3)}$$

$$i = \{0, n\},\ \Theta_0 \text{ being } \Theta_c$$

We'll represent $[(\delta C/\delta\Theta_0), (\delta C/\delta\Theta_1)...]$ as vector $\ddot{\nabla}C$ and $[\Delta\Theta_0, \Delta\Theta_1...]$ as vector $\Delta\Theta$; hence:

$$\Delta C \cong \ddot{\nabla} C * \Delta\Theta \qquad \text{(Equation 1.2.4)}$$

But suppose the following:

$$\Delta\Theta = -\alpha\ddot{\nabla} C \qquad \text{(Equation 1.2.5)}$$

Substituting it in Equation 1.2.4, we get this result:

$$\Delta C \cong -\alpha * \left(\ddot{\nabla} C\right)^2$$

Here, α being a positive number, change in loss will always be negative, and we want our loss to be negative always. So Equation 1.2.5 stands true.

Therefore by Equation 1.2.5, we get

$$\left(\Theta_i' - \Theta_i\right) = -\alpha * \delta C / \delta \Theta_i$$

$$\therefore \Theta_i' = \Theta_i - \alpha * \delta C / \delta \Theta_i \qquad \text{(Equation 1.2.6)}$$

Where Θ_i' is the new updated value for weight Θ_i. In Equation 1.2.6, the updated value of weight Θ_i' is dependent on the previous value of weight (Θ_i), the gradient ($\delta C/\delta \Theta_i$), and a positive number *α*; *α* here is one of the hyperparameters for gradient descent. It controls the performance of the algorithm. For each observation, we run this updating equation and decrease the loss while changing values of weights, eventually reaching the minima for the loss function.

The Need for Hyperparameter Optimization

In the previous section, we used a positive number *α* in Equation 1.2.6 to control the algorithm. This *α* is called the *learning rate* in the gradient descent algorithm. It controls the rate by which loss reaches its minima. Figures 1-3-1, 1-3-2, and 1-3-3 demonstrate how, as described in following figures.

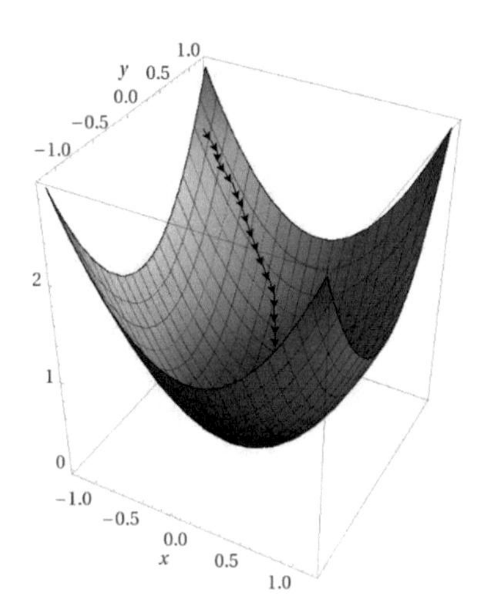

Figure 1-3-1. *Small value of α*

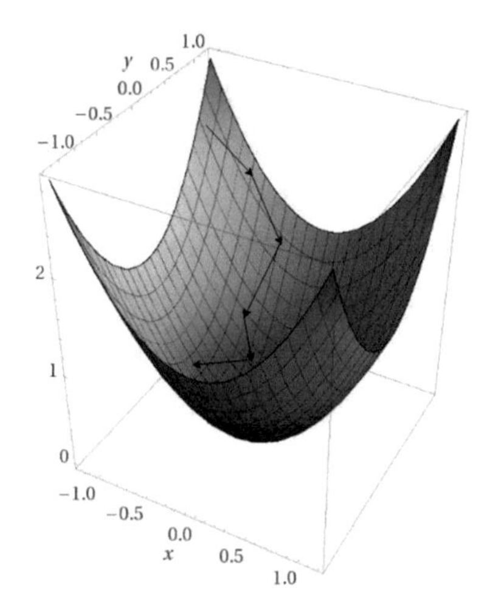

Figure 1-3-2. *Large value of α*

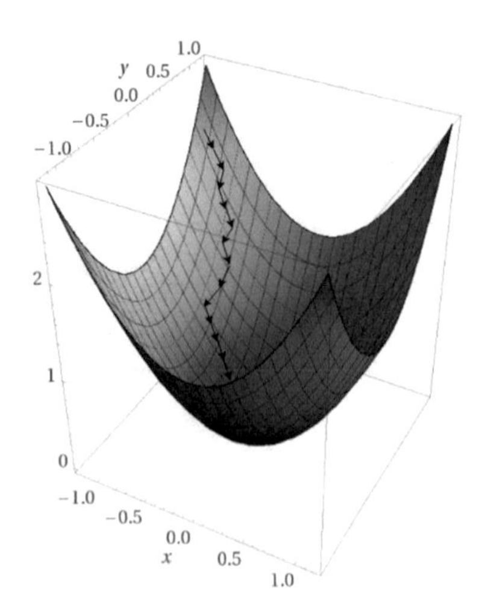

Figure 1-3-3. *Optimal value of α*

In Figure 1-3-1 the value of α is small; it will reach the convergence point, but the $\Delta\Theta$ (that is, the change in weights) will be so small that a huge number of steps would be required, hence increasing time. A large value of learning rate (α) will change loss drastically, hence overshooting and leading to divergence, as shown in Figure 1-3-2. However, if we find an optimal value of α, we'll be able to reach the convergence in less time and without overshooting, as represented in Figure 1-3-3. And that is the reason we need to tune α to its most efficient value, and this process of optimization is called *hyperparameter tuning*.

In more advanced variants of the gradient descent algorithm, we start with bigger steps (that is, a greater value of learning rate) to save time, and as we reach the convergence point, we decrease the value to avoid overshooting. But the factor by which we decrease α is now another hyperparameter. So, now you understand the importance of tuning these hyperparameters.

To tune such hyperparameters, you must have a good understanding of the algorithm and how these hyperparameters are affecting the performance. Even if you plan to use hyperparameter tuning algorithms (introduced in later chapters), it's very important to set a good starting point and ending point. This will save you a lot of time and boost the performance of your algorithm.

Algorithms and Their Hyperparameters

In this section I'll discuss some basic machine learning algorithms to help you understand how their hyperparameters work. I'll discuss these hyperparameters with scikit-learn conventions, but since they are generic, you can use them for other implementations or even self-implemented algorithms. I won't go deep into mathematics but will give you enough to get an intuition of how they affect the algorithm. In Chapter 2, we'll look at how a bad set of hyperparameters can result in a poor model, whereas a good set creates an excellent machine learning model.

K-Nearest Neighbor

The K-nearest neighbor (KNN) algorithm can be used as a supervised or unsupervised machine learning algorithm and can be applied to classification, regression, clustering, and outlier detection problems. KNN assumes similar points are in closer proximity, as depicted in Figure 1-4-1.

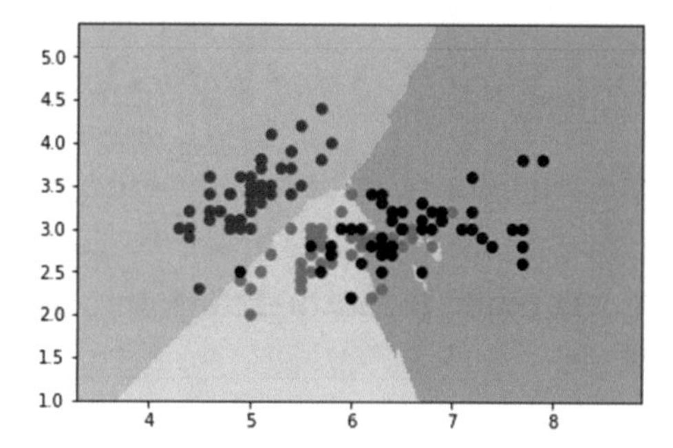

Figure 1-4-1. *A classification dataset with two dimensions when used with KNN shows the decision boundaries*

K-nearest neighbor finds the K (number of nearest points) labeled samples in the closest proximity to the point that is to be predicted. This K can be defined by the user. And the closest proximity, hence the distance, can be calculated by different metrics, such as Euclidean distance, Manhattan distance, and so on. To find these nearest points, indexing algorithms like kd-tree and ball tree are used. Let's discuss these hyperparameters.

- *K number of nearest neighbor*: We set the value of K, which is a positive integer that decides the number of labeled samples from the training dataset that are to be considered to predict the new data point. Figure 1-4-2 shows how increasing the K can result in smoother boundaries. And when k=3, boundaries are more constrained.

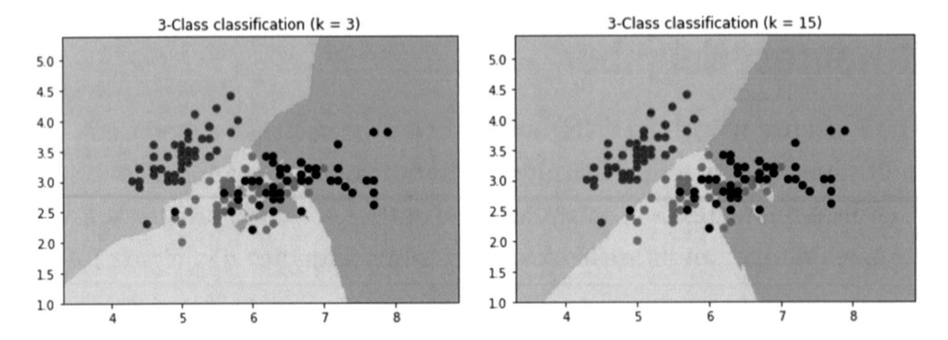

***Figure 1-4-2.** Top image with k=3 and bottom image with k=15*

- *Weights*: We can either give our nearest neighbors equal priority or decide their weights on the basis of distance from the query point; the further the point, the lesser the weightage.

- *Indexing algorithm*: Indexing algorithms are used to map the nearest points. Since brute force would result in distance computation of all the pairs of data points in a dataset, we use tree-based indexing algorithms like kd-tree and ball tree. kd-tree partitions data in cartesian axes and ball tree in nested hyper-sphere. When the number of dimensions is higher, ball tree is more efficient than kd-tree.
- *Distance metric*: A metric is to be used to calculate the distance between points. It can be Euclidean or Manhattan or higher orders of the Minkowski metric.

Support Vector Machine

Support vector machine (SVM) is a powerful algorithm that finds a hyperdimensional plane that separates distinct classes. An example is shown in Figure 1-4-3, in which we have two classes denoted by red and blue colors. The black line dividing them is our hyperplane (a line in this case since we are visualizing in two dimensions). SVM finds the hyperplane such that the margin (the distance between the two dotted lines) is maximum.

The data points lying near dotted lines are called *support vectors*. They are highly responsible for the formation of the hyperplane. We use the optimization method of Lagrange multipliers to find this hyperplane.

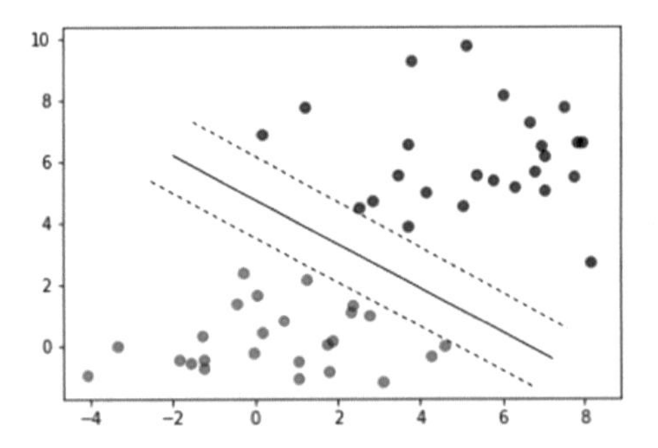

***Figure 1-4-3.** Classes separated by hyperplane*

But this was a linearly separable problem. In real life, datasets are not linearly separable. So let's take another example and see how SVM would work on the example shown in Figure 1-4-4.

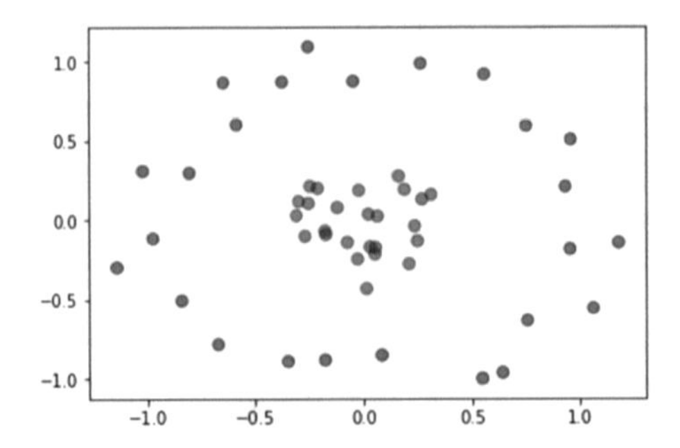

***Figure 1-4-4.** Dataset with two classes, blue and red*

Figure 1-4-4 is not linearly separable. So we project it into a higher dimension (three dimensions in this case) as shown in Figure 1-4-5 and now we can apply SVM and find the plane separating it.

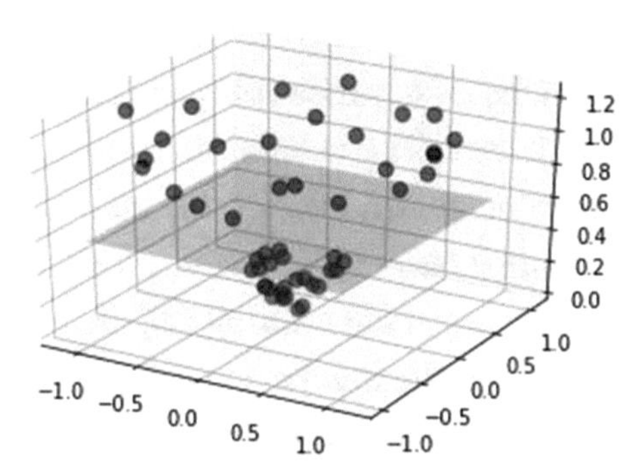

Figure 1-4-5. *Data projected into higher dimension and separated by a plane*

We need to find the correct mapping function to project data to higher dimensions. That's where different kernels come into play. Right mapping functions can be achieved from using the right kernel functions, which is one of the most crucial hyperparameters in SVMs.

Let's now discuss different hyperparameters:

- Kernel: As previously described, a kernel helps achieve the right mapping function, which is essential for SVM to perform efficiently. Finding just the kernel reduces the complexity of finding the mapping function; there's a direct mathematical relation between the mapping function and kernel function. Figure 1-4-4 is an example problem that can be solved using the radial basis function (RBF) kernel. Some of the widely used kernels are the polynomial kernel, Gaussian kernel, sigmoid kernel, and of course RBF kernel, most of them defined in the scikit-learn implementation of SVM (sklearn also allows you to define your own kernel).

- C: C is a regularization parameter. It trades off between training accuracy and the width of margin. A decrease in C results in larger margins and lower training accuracy, and vice versa.

- Gamma: Gamma (γ) defines the influence of training points. Higher value of γ means lesser influence of training point; a very high value will result in influence on training point itself. A lower value will influence more; the training points (which are support vectors) will influence more of the training set, hence extremely lower values will be ineffective in catching the complexity of dataset and the resulting hyperplane will be equivalent to a linear hyperplane separating two classes (by their density regions).

  This establishes an interesting relation between hyperparameters C and Gamma. Generally, we search values of Gamma and C on a logarithmic grid of 10^{-3} to 10^{3}.

- Degree: This hyperparameter is used only in polynomial kernels; a higher degree means a more flexible decision boundary. Degree 1 would result in a linear kernel.

Decision Tree

Decision tree is similar to a bunch of if-else statements, a simple yet elegant algorithm with a very intuitive visualization. It's really easy to understand what's going inside, unlike neural networks. Moreover, little or no data preprocessing is required.

As the name suggests, it's a tree, so it starts with a root node, which is one of the features. Based on the value of that feature for our data point, we select the next node of the tree. This goes on until we reach the leaf and thus the prediction value.

Creating this tree is a little bit complicated; various different algorithms are used to select which feature goes on the top, which goes second, and so on. Some of the algorithms which calculate the importance of features are Gini index, information gain, and chi-square. Selecting this algorithm can be considered as one of the important hyperparameters in the decision tree algorithm.

Let's take the example of a classical Iris dataset. Here the goal is to classify the three species of Iris flowers, Setosa, Versicolor, and Virginica, based on four features, sepal length, sepal width, petal length, and petal width.

As I said earlier, visualization of a decision tree is very easy; sklearn provides a function, `tree.plot_tree()`, where you just have to input your trained classifier and it will plot the tree (Figure 1-4-6).

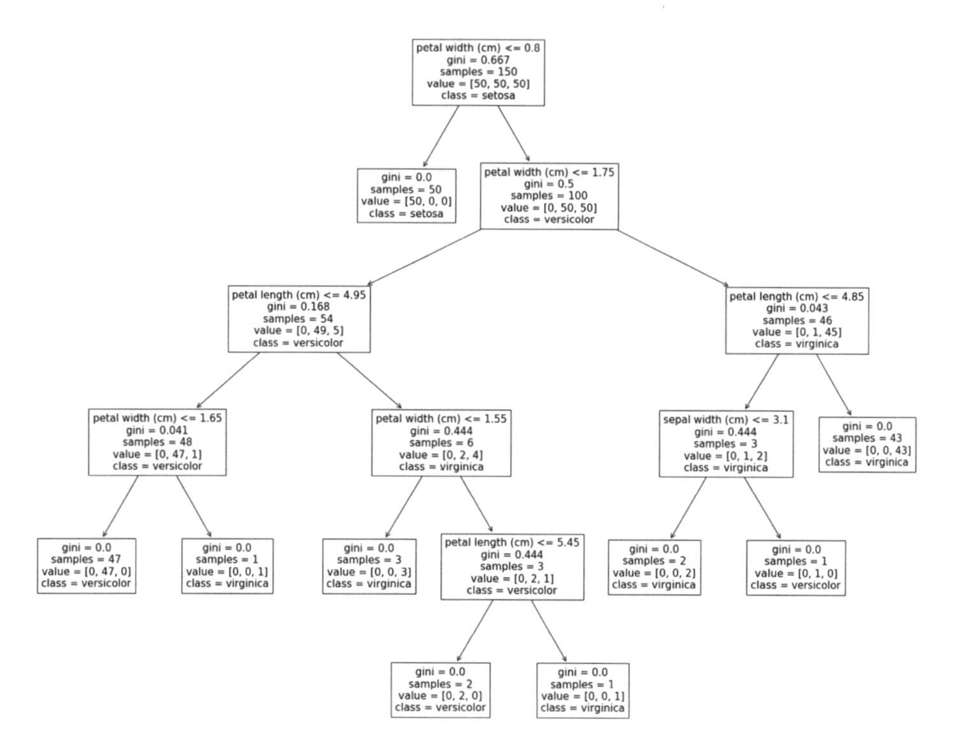

Figure 1-4-6. *A decision tree classifier trained on the Iris dataset*

As you can see in Figure 1-4-6, there are gini indexes for all the nodes, based on which features are placed in the tree. As we move down the tree, the value of gini index decreases. On top of nodes, a condition is specified; if it's true, the data point goes to the right child, and if it's false, the data point goes to left child. The value of samples tells us the number of samples lying in the true and false condition of the parent node.

One of the problems we face with decision trees is that when the tree grows complex, there is a huge chance the model will overfit over training data. Some of the hyperparameters can help in reducing this complexity. To solve this problem, we can prune the tree, using hyperparameters such as maximum depth of tree and minimum number of samples at the leaf node to help in pruning.

Here are the hyperparameters:

- Algorithm: As previously mentioned, this algorithm decides the priority of features and hence their order in the tree structure.
- Depth of Tree: This defines the layer of depth. This can certainly affect both structural and time complexity of the tree. We can remove unimportant nodes and reduce depth.
- Minimum Sample Split: This is an integer value that defines the minimum number of samples required to split an internal node. In Figure 1-4-6, if we would have chosen 101, the tree would have stopped after the second layer.
- Minimum Sample Leaf: This defines the minimum number of samples at the leaf. This hyperparameter can help reduce overfitting by reducing the depth of the tree.

Neural Networks

A basic *neural network* is made of nodes and layers of nodes, and these nodes are nothing but the output of the previous layer multiplying with weights (see Appendix II for more details). We call weights and biases in this context *parameters* (since they are decided by a modeling algorithm based on the dataset) and we call the number of nodes, number of layers, and so on *hyperparameters* (since we intervene to decide them).

Defining the architecture of a neural network is one of the most challenging tasks faced by deep learning practitioners today. The architecture can't be discovered by brute force because the time complexity of neural networks is very high and trying out all the combinations of hyperparameters is not possible. So, creating a neural network architecture is more of an art, relying on logic and more advanced hyperparameter tuning algorithms.

A vast number of different hyperparameters are possible in neural networks, so we'll discuss a few of them here:

- Number of Layers: Adding layers increases the depth of the neural network, and also the ability to learn more complex features.
- Number of Nodes: The number of nodes varies as per the layers, but the number of nodes in the first hidden layer and the last hidden layer must be equal to the number of input features and classes to predict, respectively. For the hidden layer, by convention we use the number of nodes in exponents of 2, meaning 2, 4, 8, 16, 32, 64, 128, 256, 512, 1024, and so on. This is because hardware performs more efficiently when numbers are stored in powers of two, though there is no proof that this is the most optimal way of selecting these kind of hyperparameters.

- Batch Size: If we take out a subsample of the dataset, it should represent the properties of the whole dataset. This batch can be used to calculate the gradient and update the weights. And we iterate over all the subsamples until we cover the whole dataset. The idea is to save memory space. But you need to choose the optimal value of the size of the subset, because a lesser batch size would cause more fluctuations while reaching the minima, and a greater value can cause memory errors.
- Activation Function: Activation functions are used to introduce a nonlinearity on each node. Few things we need to make sure while deciding activation functions are, they are to be used on thousands and millions of nodes, and back propagation uses their derivatives, so both the function and its derivative should be less computationally complex. Some of the widely used activations are ReLU, Sigmoid, and Leaky ReLU.
- Loss Function: Loss function is chosen on the basis of output, whether it's a binary classification, multi-class classification, regression, and so forth. There are also other factors. For example, using sigmoid activation on the last layer and quadratic loss function can result in learning slow down. So things like these are need to be taken care of. And there are internal hyperparameters for loss function as well which can be tuned.

- Optimizer: In the "Understanding Hyperparameters" section we discussed an optimization method, gradient descent. There are other, more advanced optimization methods, like Adagrad, Adam Optimizer, and so on, and these optimizers also contain various hyperparameters that affect the overall optimization.

There are many more hyperparameters in neural networks, such as batch normalization, dropout, and so on. And every few days these variables are increasing with the advance in technology.

This section was intended to give you an idea of what hyperparameters are and how they work. As we proceed we'll be discussing more of these algorithms and their hyperparameters while finding their optimal values.

Distribution of Possible Hyperparameter Values

The value of a hyperparameter can vary based on its functioning. For algorithms of the likes of grid search (discussed in Chapter 2), we iterate over certain permutations of hyperparameters. But most of the hyperparameter optimization algorithms pick variables at random. These random values can belong to a certain type of distribution. For example, we saw in the "Neural Networks" section that we choose the value for the number of nodes in a layer to be an exponent of 2. So, we consider the set {2, 4, 8, 16, 32, 64, 128, 256, 512...} to be a distribution.

Tip Seeing Theory (`https://seeing-theory.brown.edu/`) is a great website with interactive visualizations of probability and statistics. Check out Chapter 3 of this website for visualizations of probability distributions.

The likelihood of a value that a random variable can assume is defined by probabilistic distribution. Suppose we have to pick a single random value for a hyperparameter; the underlying distribution can be either of the following:

- Discrete probabilistic distribution
- Continuous probabilistic distribution

There are different types of probabilistic distributions for both discrete and continuous variables. But before we delve into those types, let's look at what discrete and continuous variables are.

Discrete Variables

A set of values where each value has some positive finite distance to the next value is *discrete distribution*. Discrete variables can be either finite values or infinite values depending upon range. Figure 1-5-1 shows both finite discrete and infinite discrete distributions.

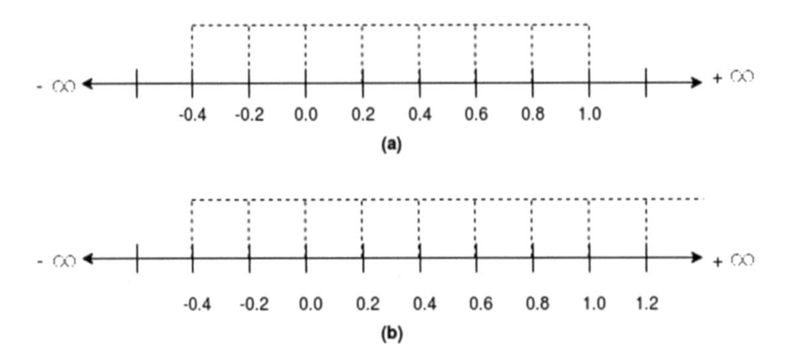

Figure 1-5-1. *Two distributions, (a) finite discrete and (b) infinite discrete*

A finite discrete value lies between two finite real numbers, as shown in Figure 1-5-1(a), where the value lies in the range [−0.4, 1.0] with a difference of 0.2. However, an infinite discrete value can go up to infinity, each value still maintaining a finite positive distance to the next value and the previous value. In Figure 1-5-1(b), the value lies between $[-0.4, +\infty)$ while the difference is still 0.2.

Note that the difference does not necessarily have to be same. It can be exponential, incremental, and so forth. The example in Figure 1-5-2 shows a uniform distribution.

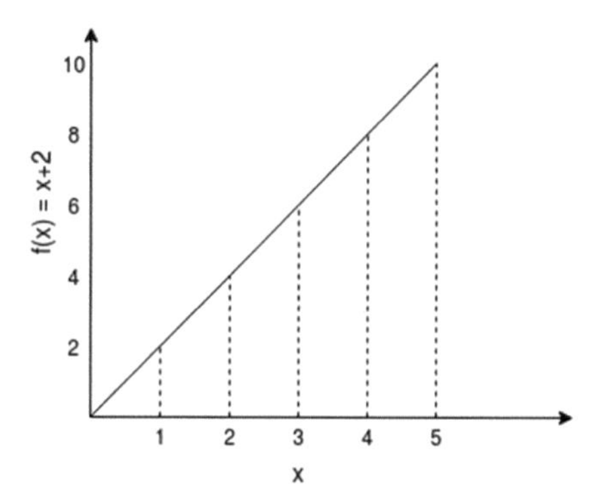

Figure 1-5-2. *An example of finite distribution, $f(x) = x + 2$ such that $x \in [0, 5]$ with a difference of 1 between each value of x*

Recall from the "Support Vector Machine" section that possible values for the hyperparameter *kernel* can be *rbf, sigmoid, linear, polynomial,* and so on. There would be a finite set of values for *'kernel'*. This can be considered an example of a *finite discrete value.* We can write these hyperparameters as follows:

- *rbf* $\leftrightarrow$ 1
- *sigmoid* $\leftrightarrow$ 2
- *linear* $\leftrightarrow$ 3
- and so on

In the same algorithm we have another hyperparameter, *'degree'*, whose value can be any possible integer. This is an example of an *infinite discrete value.* However, this does not mean that we are going to search this value from $-\infty$ to $+\infty$. The results would saturate on extreme value and it is also not practically possible to search in an infinite space. So we'll use a huge range (huge is relative; 10 can be huge in some cases, while 10^{100} can be huge in others) to contain the distribution.

Continuous Variables

Continuous distribution is a set of infinite possible values lying between two real numbers, as depicted in Figure 1-5-3.

Figure 1-5-3. *Infinite numbers lie between 1.0 and 2.0*

Again taking an example from the "Support Vector Machine" section, we have hyperparameters like 'C' and 'gamma' in SVM, the values of which lie on a continuous distribution; that is, we can have infinite possible values between a range.

Probabilistic Distributions

There are infinite possible continuous and discrete probabilistic distribution functions to sample random values, so we'll narrow the scope by discussing a few commonly used in practice. Probability is always calculated between a range, such as in the example Gaussian distribution shown in Figure 1-5-4.

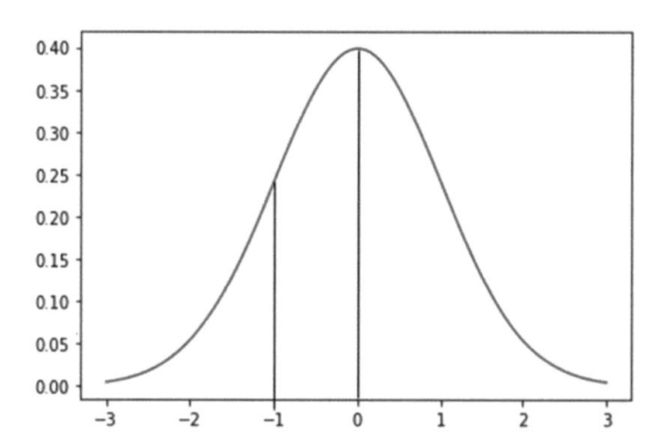

Figure 1-5-4. *A Gaussian distribution with mean 0 and standard deviation 1*

In Figure 1-5-4, if we want to sample a value for variable x, there are infinite values that x can assume. The probability for $p(x)$ to be a specific value would be 0. So we talk in terms of *probability density function*, which are probabilities in a range. For example $p(x < 0) = 0.5$. Since it's half of the area from the whole distribution. Similarly, probability density in the range -1 to 0 can be calculated by using the area, which can be calculated using integration along the continuous curve.

This explains why we need to get the probability density of a range instead of a value. Next we'll use a module named *scipy. stats* to sample values for variable x for purposes of discussing some commonly used distributions.

Uniform Distribution

In *uniform distribution*, probability density remains the same across ranges if the width is the same. Figure 1-5-5 shows an example.

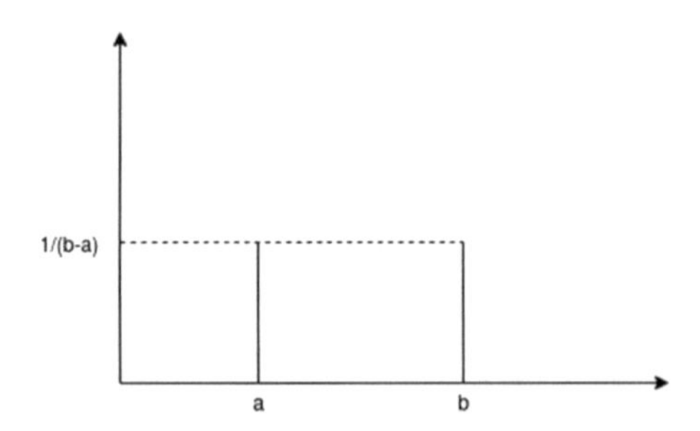

Figure 1-5-5. *A uniform distribution between a and b*

The area under the curve is 1. So we can calculate the height as 1/($b - a$), where b is the upper limit and a is the lower limit. Let's use *scipy.stats* to sample values from the uniform distribution:

```
from scipy.stats import uniform
import seaborn as sns

n = 10000
start = 10
width = 10
data = uniform.rvs(size=n, loc = start, scale=width)
ax = sns.distplot(data,
                  bins=100,
                  hist_kws={'alpha':0.8})
ax.set(xlabel='Uniform Distribution ', ylabel="Frequency")
```

Figure 1-5-6 shows the histogram for the uniform distribution, where a=10 and b=20. I have sampled 10,000 random values between two numbers.

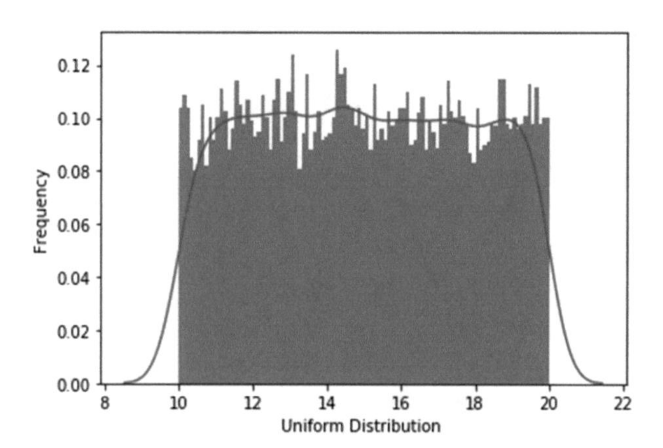

Figure 1-5-6. *Histogram for uniform distribution*

Gaussian Distribution

A *Gaussian distribution* (or *normal distribution*) is one of the most commonly observed distributions in nature. Most of the machine learning algorithms assume the dataset to be Gaussian; that is, less number of outliers and most data concentrated in clusters. Lesser frequency on extremities and higher frequency on the mean.

For a mean (μ) of 0 and standard deviation (σ) of 1, you can see a Gaussian distribution in Figure 1-5-4. Note that on the x axis, if you mark $\mu - \sigma$ and $\mu + \sigma$, as in Figure 1-5-7, you'll find that it covers approximately 68% of the area. Similarly, $\mu - 2\sigma$ and $\mu + 2\sigma$ covers around 95% of the area and $\mu - 3\sigma$ and $\mu + 3\sigma$ covers 99.7 % of the area.

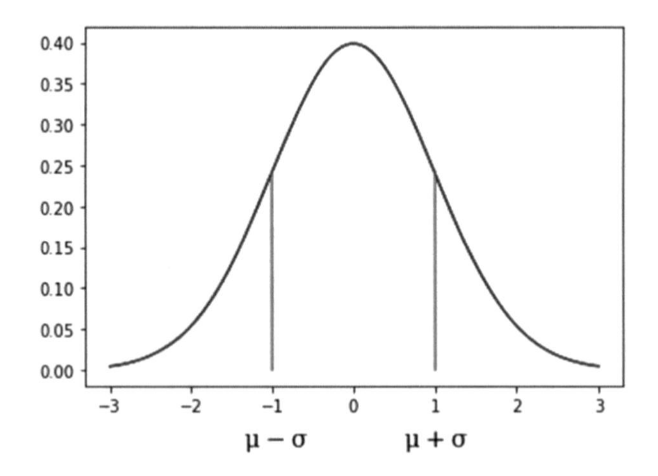

Figure 1-5-7. *A Gaussian distribution with mean 0 and standard deviation 1*

Here's the code to sample random values from Gaussian data:

```
from scipy.stats import norm
import seaborn as sns

mean = 0
std_dev = 1
data = norm.rvs(size=10000,loc=0,scale=1)
ax = sns.distplot(data,
                  bins=100,
                  hist_kws={'alpha':0.8})
ax.set(xlabel='Gaussian Distribution ', ylabel="Frequency")
```

Figure 1-5-8 shows a histogram plotted from random values sampled from a Gaussian distribution.

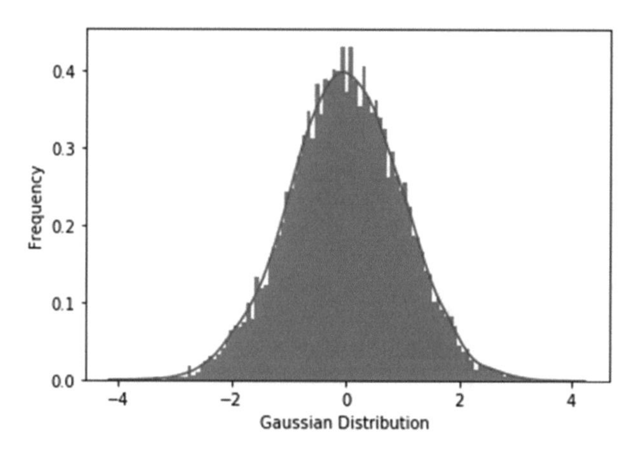

Figure 1-5-8. *Histogram for Gaussian distribution*

Exponential Distribution

Another important distribution is *exponential distribution*. As the name suggests, values increase exponentially. There's a parameter rate (λ) which controls the slope of distribution.

```
from scipy.stats import expon
import seaborn as sns

loc = 10
lambda_inverse = 1
data = expon.rvs(size=10000,loc=loc,scale=lambda_inverse)
ax = sns.distplot(data,
                  bins=100,
                  hist_kws={'alpha':0.8})
ax.set(xlabel='Exponential Distribution', ylabel="Frequency")
```

Figure 1-5-9 shows a histogram for when a random value is selected from a lognormal distribution which can be plotted using the above code.

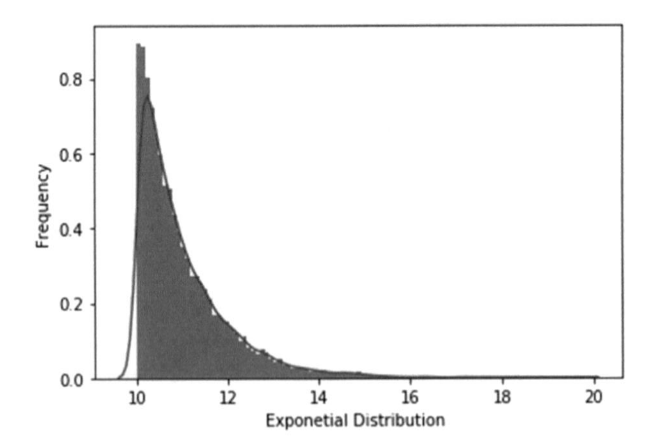

Figure 1-5-9. *Histogram for an exponential distribution*

We'll encounter a lot more distributions in subsequent chapters as we use them. We'll see which hyperparameters are suited for which distributions and why. Now that you have an understanding of hyperparameters, in Chapter 2 we'll explore some basic hyperparameter tuning methods.

CHAPTER 2

Hyperparameter Optimization Using Scikit-Learn

In the previous chapter, you learned what hyperparameters are and how they affect the performance of an algorithm. Now that you know how important it is to tune hyperparameters, this chapter introduces you to some simple yet powerful uses of algorithms implemented in the scikit-learn library for hyperparameter optimization. Scikit-learn is one of the most widely used open source libraries for machine learning practices. It's simple to use and really effective in predictive analysis.

Changing Hyperparameters

You know from Chapter 1 how support vector machine (SVM) works. You'll now see how changing two of the hyperparameters—C, the regularization factor, and gamma, the kernel coefficient—affects the results while the kernel is fixed (RBF) on the Titanic dataset (Dataset explained in Appendix I).

Figure 2-1(a) shows the comparison between gamma and C; the lighter color in the heat map represents higher accuracy. We see that with higher values of C (10^11) and lower values of gamma (10^-8), we get more test accuracy, and with lower values of C (10^4) and relatively higher values of

T. Agrawal, *Hyperparameter Optimization in Machine Learning*,
https://doi.org/10.1007/978-1-4842-6579-6_2

gamma (10^-3), we get comparative accuracy. The graph in Figure 2-1(b) shows that as the value of gamma increases, keeping the C constant at 1, the difference between train accuracy (blue line) and test accuracy (orange line) after a certain point keeps on increasing, resulting in overfitting of the model, which proves that we need to regularize the model by decreasing C.

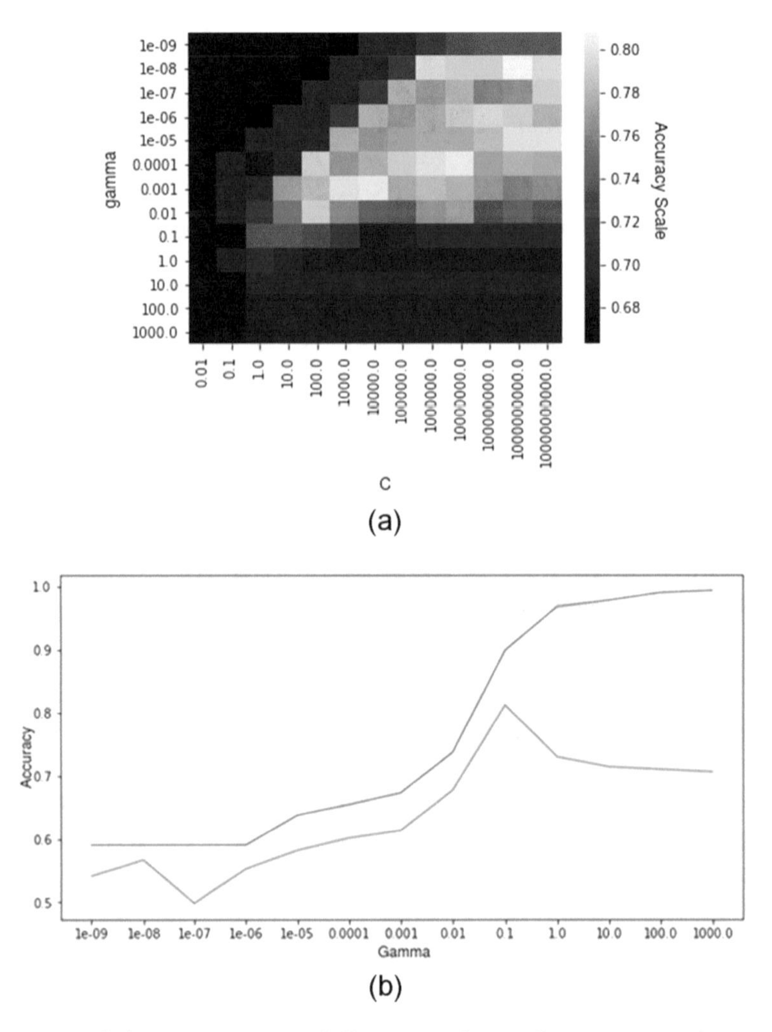

***Figure 2-1.** (a) Heatmap on different values of gamma and C. (b) Changing accuracy as gamma increases*

Grid Search

Perhaps the most brute-force approach for finding the most optimized set of hyperparameters is to train the dataset on each possible set. This approach, called *grid search,* is the most certain way of finding the best set of hyperparameters, but it also has its disadvantages. Figure 2-1-1 depicts a grid going through all possible combinations of parameters 1 and 2.

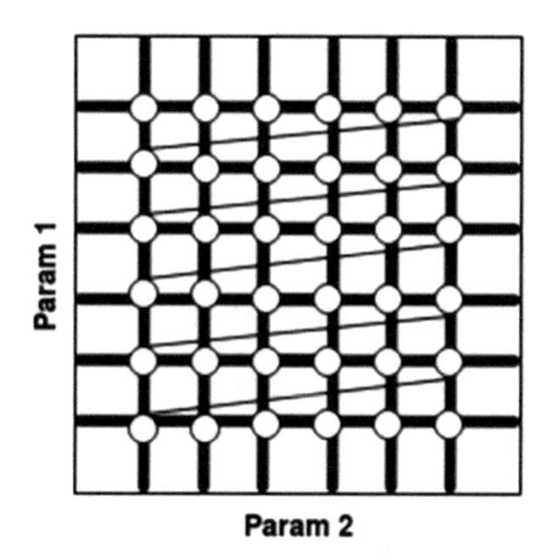

Figure 2-1-1. *A grid search going through each possible combination of two hyperparameters*

Suppose you have ten algorithms with five hyperparameters each, with four possible values for each hyperparameter, and a huge dataset that takes 1 minute to train on one algorithm on one set of hyperparameters. This scenario would take around a week to discover your best hyperparameter. But for less number of hyperparameters and smaller training time we can go with grid search. Now we'll build this simple algorithm in Python and test it on the example shown in Figure 2-1-1.

The value of a hyperparameter can vary on either a continuous distribution or a discrete distribution. If the value is discrete, there is a finite number of possible values in a range. However, in a continuous distribution, there are infinite possible values, as we saw in Chapter 1. Next we'll be tuning C and gamma, both of which have a continuous distribution between a range. But to make a grid, we'll make an even distribution on a log scale. Grid search does not take random variables; to make a grid, it needs specific values.

Let's make a grid of hyperparameters C and gamma:

```
c = 0.001
gamma = 1e-10
param_grid = {
              "C": [c*(10**i) for i in range(1,14)],
              "gamma": [gamma*(10**i) for i in range(1,14)]
             }
```

Note I have used a preprocessed dataset (X_train, y_train, X_test, y_test), the Titanic dataset. Refer to Appendix I to view all the preprocessing methods.

We'll first make a function to break this grid into a list of all possible sets of hyperparameters, make_sets():

```
from itertools import product

def make_sets(grid):
   sets = list()
   all_hps_vals = [lst for lst in param_grid.values()]
   hp_keys = [hp for hp in param_grid.keys()]
   val_sets = product(*all_hps_vals)
   for val in val_sets:
       hp_set = dict()
       for idx, hp_key in enumerate(hp_keys):
           hp_set[hp_key] = val[idx]
       sets.append(hp_set)
   return sets
```

```
>>> make_sets(param_grid)
[{'C':0.01, 'gamma':1e-09},
 {'C':0.01, 'gamma':1e-08},
 {'C':0.01, 'gamma':1e-07},
 ...
 ...
 {'C':10000000000.0, 'gamma':1000.0}
]
```

Now we'll make another function, grid_search(), to fit all the sets on the machine learning algorithm:

```
def grid_search(clf, grid, X_train, y_train, X_test, y_test):
   all_sets = make_sets(grid)
   logs = list()
   best_hp_set = {
      "best_test_score": 0.0
   }
   for hp_set in all_sets:
       log = dict()
       model = clf(**hp_set)
       model.fit(X_train, y_train)
       train_score = model.score(X_train, y_train)
       test_score = model.score(X_test, y_test)

       log["hp"] = hp_set
       log["train_score"] = train_score
       log["test_score"] = test_score

       if best_hp_set["best_test_score"]<test_score:
           best_hp_set["best_test_score"] = test_score
           best_hp_set["hp_set"] = hp_set

       logs.append(log)
```

```
    return logs, best_hp_set

>>> from sklearn.model_selection import train_test_split
>>> from sklearn.svm import SVC

# train test split dataset.
# X and y are the pre-processed features and labels respectively.
>>> X_train, y_train, X_test, y_test = train_test_split(X, y)

>>> logs, best = grid_search(SVC, param_grid, X_train, y_train,
X_test, y_test)
```

The `grid_search()` function that we just defined takes the following inputs: classifier, parameter_grid, and dataset. From the `make_sets()` function, `grid_search()` creates all combinations of hyperparameters and trains the model on all of them. Then we save the train and test scores in a dictionary and search for best results.

Note that for the sake of simplicity of code, I did not *cross-validate* while training. To actually evaluate each set of hyperparameters, we must use a validation set and save the test set for later, so that we can evaluate the model on an independent set. However, instead of splitting the training set into training and validation sets (since in datasets like Titanic, we have only around 700 datapoints), we cross-validate, saving the precious training data unaltered. Cross-validation also prevents overfitting on the validation set.

We'll now see how to use the `GridSearchCV()` function provided by scikit-learn to split the training set in a K-fold cross-validation:

```
>>> from sklearn.model_selection import GridSearchCV

>>> clf = GridSearchCV(SVC(), param_grid, cv=3)

>>> # X_train and y_train being datapoints from titanic dataset.
>>> # titanic dataset is used for sake of presenting this example.
>>> clf.fit(X_train, y_train)
```

```
>>> clf.best_estimator_
SVC(C=100000.0, break_ties=False, cache_size=200, class_
weight=None, coef0=0.0,
    decision_function_shape='ovr', degree=3, gamma=0.0001,
    kernel="rbf",
    max_iter=-1, probability=False, random_state=None,
    shrinking=True,
    tol=0.001, verbose=False)
```

In Figure 2-1-2, we can see that the accuracy score varies with iterations (for 169 combinations).

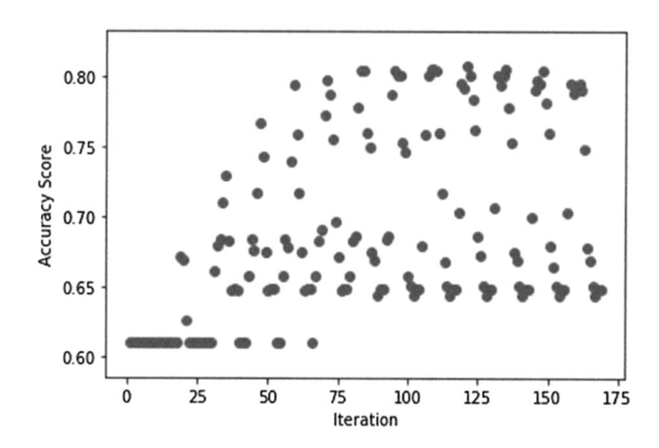

***Figure 2-1-2.** Plot of iteration vs. accuracy score*

Figure 2-1-2 exhibits no clear pattern since it's an exhaustive search method; we are trying all possible combinations. You can define the `GridSearchCV()` function and pass the algorithm, parameter grid, and number of folds for cross-validation. All the other methods like 'fit()', 'score()' and so forth are same. You can use the method `best_estimator_` to get the best value of hyperparameters. I tuned for 'gamma' and 'C' the same as our scratch implementation, keeping the rest of the hyperparameters constant.

Random Search

Grid search eventually finds the near optimal set of hyperparameters, but its time and resource consumption is high. Another method, *random search,* consumes less time and resources. It randomly picks hyperparameters, makes a set, and trains the model on it. This method may not find the best set, but there are higher chances of finding a near best set saving a huge amount of time.

Unlike grid search, instead of spending a large amount of time on unpromising candidates, random search jumps to random hyperparameters, and even though it does not learn from its past results, it usually delivers satisfactory results. In random search, we define the *number of trials,* which is the number of sets of hyperparameters to be tried.

Let's see how random search can be better than grid search by exploring the example shown in Figure 2-2-1.

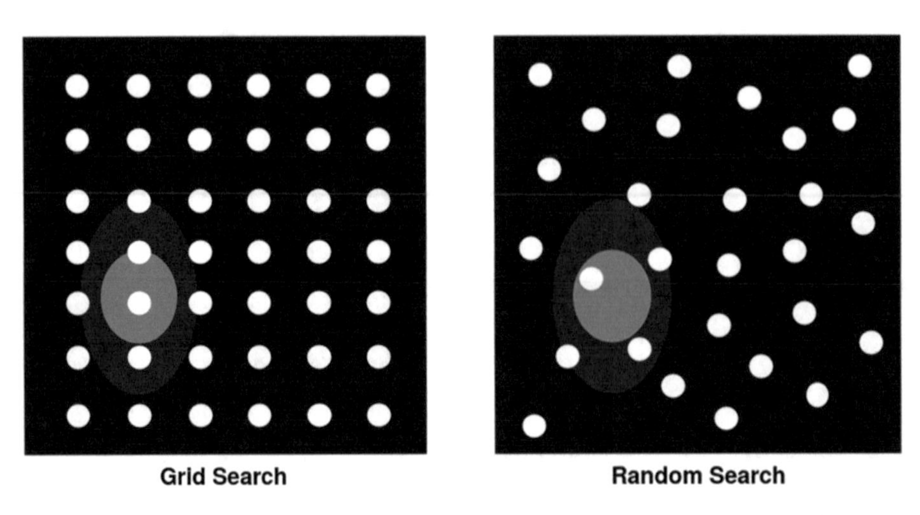

Figure 2-2-1. *Comparing grid search (left) to random search (right)*

In both images in Figure 2-2-1, 'x' and 'y' axis represents two hyperparameters and the background represents increasing accuracy as the color gets lighter. In the case of grid search, shown on the left, if we start searching from the top-left corner, along the grid, our search will take a considerable amount of time to reach the higher-accuracy region. In the case of random search, shown on the right, because we are randomly searching for hyperparameters, we have better chances of reaching higher accuracy earlier than with grid search. And as soon as the defined number of trials are over, we'll select the best set of hyperparameters available, hoping it's at least near the best set.

The following are two main benefits of using random search over grid search:

- The number of trials is defined and is independent of the total number of combinations.
- Since the number of trials is defined, even if the number of noncontributing parameters is increased, the time efficiency of the algorithm isn't affected.

In random search, since we select hyperparameters randomly out of the search range, we can use the `random` library in Python, `random.randint(a, b)` (which gives a random integer between integers a and b) for discrete hyperparameters and `random.random()` (which gives a random float number between 0 and 1, where 1 is exclusive) for continuous or functions from numpy or scipy.stats which gives different types of distributions like uniform, lognormal, exponential, and so forth.

Alternatively, as shown next, we can create a bigger grid for hyperparameters with continuous distribution—like really large, since it won't increase the number of trials (which we are going to define)—enabling the algorithm to choose hyperparameters from a bigger sample.

```
import random
import numpy as np

def loguniform(low=0, high=1, size=100, base=10):
        # function creates a log uniform distribution with
        # random values.
        return np.power(base, np.random.uniform(low, high, size))
```

```
param_grid = {
              "gamma": loguniform(low=-10, high=4, base=10),
              "C": loguniform(low=-3, high=11, base=10)
             }

def get_random_hp_set(grid):
      # function chooses a random value for each from grid
      hp_set = dict()
      for key, param in grid.items():
            hp_set[key] = np.random.choice(param)
      return hp_set

def random_search(clf, grid, n_iterations, X_train, y_train,
X_test, y_test):
      # defining function for random search
      logs = list()
      best_hp_set = {
       "best_test_score": 0.0
      }

      for iteration in range(n_iterations):
            log = dict()

            # selecting the set of hyperparameters from
            function defined
            # for random search.
            hp_set = get_random_hp_set(grid)
            # print(hp_set)
            model = clf(**hp_set)
            model.fit(X_train, y_train)
            train_score = model.score(X_train, y_train)
            test_score = model.score(X_test, y_test)
```

```
        log["hp"] = hp_set
        log["train_score"] = train_score
        log["test_score"] = test_score

        if best_hp_set["best_test_score"]<test_score:
            best_hp_set["best_test_score"] = test_score
            best_hp_set["hp_set"] = hp_set

        logs.append(log)

    return logs, best_hp_set

>>> X_train, y_train, X_test, y_test = train_test_split(X, y)

>>> logs, best = random_search(SVC, param_grid, 20, X_train,
y_train, X_test, y_test)
```

And hence we would get at least the near best set of hyperparameters from random search in lesser iterations.

Again scikit-learn provides us with a cross-validating function, `RandomizedSearchCV()`. Let's see how it works:

```
>>> from sklearn.model_selection import RandomizedSearchCV

>>> # just like our function RandomizedSearchCV also has
argument 'n_itern'
>>> clf = RandomizedSearchCV(SVC(), param_grid, n_iter=20, cv=3)

>>> clf.fit(X_train, y_train)

>>> clf.best_estimator_
SVC(C=1000000000.0, break_ties=False, cache_size=200, class_
weight=None,
    coef0=0.0, decision_function_shape="ovr", degree=3,
    gamma=1e-05,
    kernel='rbf', max_iter=-1, probability=False, random_
    state=None,
    shrinking=True, tol=0.001, verbose=False)
```

As shown in Figure 2-2-2, random search reached comparable accuracy in just 20 iterations.

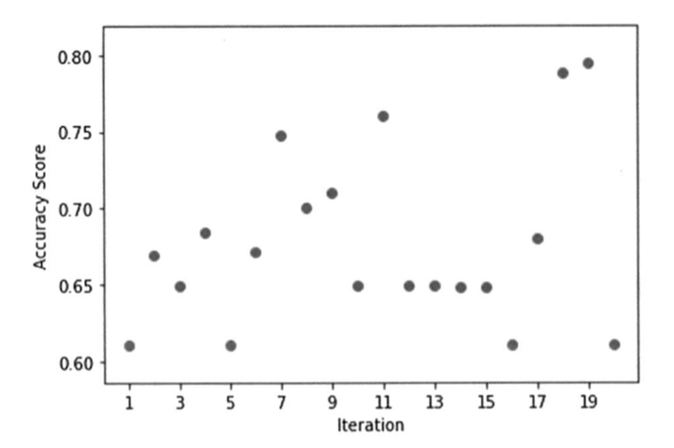

Figure 2-2-2. *Iteration vs. accuracy plot for random search*

Both `GridSearchCV` and `RandomizedSearchCV` have another useful argument, `scoring`; if set to its default value, the machine learning algorithm's scoring method is used, which is usually `'accuracy'`. However, we can provide it with any of the scoring methods we find fit for our work, like `'roc_auc'`, `'f1'`, `'precision'`, `'recall'`, and so forth.

Parallel Hyperparameter Optimization

Both grid search and random search are brute-force methods; they do not depend on the previous results to select the next set of hyperparameters. And we can use this to our advantage by processing the trials in parallel. The scikit-learn implementation of these algorithms provides a parameter, '`n_job`', that can be set to '`-1`' to specify that all cores of the local machine should be used.

In this section we'll see how we can distribute algorithms like grid search on a high-performance computing (HPC) cluster. An HPC cluster is simply a bunch of high-end computers (called *nodes*) that are configured so that they can be in constant communication with the help of a fast interconnect, providing computing power similar to that of a supercomputer. An HPC cluster consists of the following components, depicted in Figure 2-3-1:

- *Login node*: From here we can remotely access the cluster using our machine, through secure protocols such as Secure Shell (SSH). It is used to upload and execute the code.
- *Data transfer node*: Again, secure protocols such as SSH are used with commands like `scp` and `rsync` to transfer large amounts of data from our machine to the HCP cluster. The transfer is secured by an encryption tunnel.
- *Computer nodes*: There are different types of computer nodes, including regular ones with hardware specifications similar to those of our personal machines, "fat" computer nodes with huge amounts of data storage (in terabytes), and high-end nodes consisting of GPUs and CPUs. These nodes together help in computation.
- *Infinite Band (IB) switch*: This switch enables fast communication between nodes, with high throughput and very low latency.
- *Storage*: Here large files can be stored and can be transferred via the data transfer node.

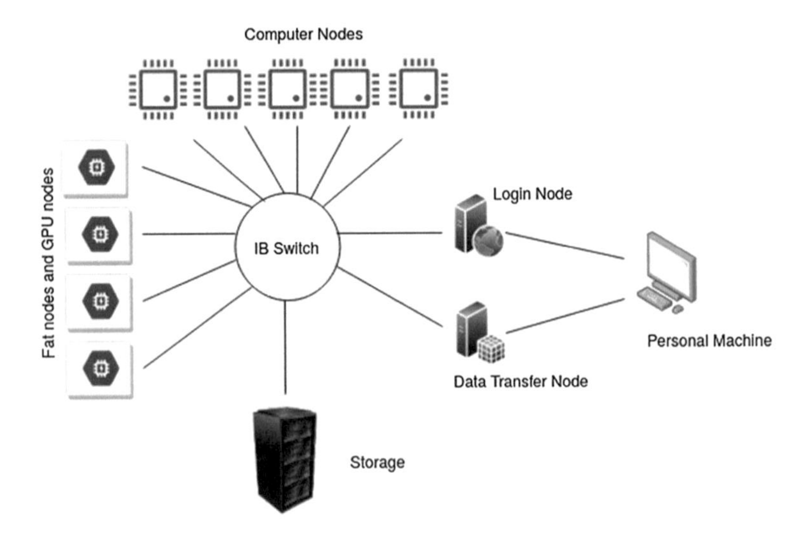

Figure 2-3-1. *HPC cluster*

However, working with high-performance computing clusters is not as simple as setting 'n_job = -1', and using clusters to their maximum potential is even difficult. We'll be using the Slurm workload manager, an open source job scheduler for computer clusters for queuing, and ipyparallel for parallel computations over multiple threads.

You can use the following code for HPC and Amazon EC2 clusters with minor modifications based on the respective requirements.

Now we'll again optimize SVM's hyperparameters C and gamma on the MNIST dataset (refer to Appendix II to know more about MNIST).

The following code is inspired by the work[1] of Dr. Hugues-Yanis Amanieu, a data scientist in production engineering at Leclanché.

First we'll log in to the login node, using SSH:

```
$ ssh username@ip
```

[1] http://www.hyamani.eu/2018/05/20/parallel-super-computing-with-scikit-learn/

Now that we are logged in to the HPC cluster, we'll create a virtual environment:

```
$ virtualenv hpc_tuning
$ source hpc_tuning/bin/activate
```

And install all the dependencies, slurm, slurm-client, ipyparallel, joblib, ipython, and of course scikit-learn in your virtual environment 'tuning'. We can alternatively download and install them through Anaconda as well (in a conda virtual environment).

Next we'll write a shell script ('launch.sh') in which we'll use Slurm to schedule the jobs:

```
#this will activate the virtualenv
source hpc_tuning/bin/activate

#creating a new job profile name for ipython which slurm will use
profile=job_${JOB_NAME}

#creates an config file for ipython
ipython profile create ${profile}

#starts ip controller
ipcontroller --ip="*" --profile=${profile} &
sleep 10

#srun runs ipengine on all the cores
srun ipengine --profile=${profile} --location=$(hostname) &
sleep 25

#execute the python file, where we'll define the grid search to
#distribute on cluster
python $1 -p ${profile}
```

Let's explore this code in detail.

- First we activate a virtual environment where everything is installed from scratch, so that we don't face any dependencies issues.
- After that, we assign a variable `profile`, where we define a string that will be the name/identity of our job.
- Once the ID is created, we use the `ipython` command to create a profile, which initializes a folder containing configuration information.
- Now we have a controller and engines. The controller schedules and queues the jobs, and engines compute the data and store the results. We are using ipython for communication between the controller and the engines.
- The `ipcontroller` command starts the controller and the `ipengine` command starts the engines. While starting 'ipcontroller', we need the controller to listen to all the engines, and the argument `--ip="*"` allows the controller to listen on all interfaces.
- For engines to connect with the controller, we use `ipengine` with the argument `--location=="ip"`, where we provide the IP address of the controller.
- `ipcontroller` creates a file named 'ipcontroller-engine.json' that needs to be copied to all the engines; however, in our case we assume the engines and the controller share the same file system, in which case engines will automatically find the location of the 'json' file.

Note in the preceding code that we are using `srun` to run `ipengine`; `srun` is a Slurm command that runs `ipengine` on all the available cores. Finally, we run the Python script containing the scikit-learn code, and an argument profile name is passed.

Let's write the Python script ('python_script.py'):

```
import argparse
import os
import sys
import time
import pandas as pd
from sklearn.externals.joblib import parallel_backend
from sklearn.externals.joblib import register_parallel_backend
from sklearn.externals.joblib import cpu_count
from sklearn.datasets import load_digits
from sklearn.model_selection import GridSearchCV
from sklearn.model_selection import train_test_split
from sklearn.svm import SVC
from ipyparallel import Client
from ipyparallel.joblib import IPythonParallelBackend

# append file dir path to sys path, so imports from custom
function would # work
FILE_DIR = os.path.dirname(os.path.abspath(__file__))
sys.path.append(FILE_DIR)

# argparser to take profile name as argument from our shell
script
parser = argparse.ArgumentParser()
parser.add_argument("-p", "--profile", default="job_hp_test",
                    help="Name of IPython profile to use")
args = parser.parse_args()
profile = args.profile
```

```
# counts total number of available cores
print(cpu_count())

# Create a Client instance providing the name of profile
created on shell script
c = Client(profile=profile)

# Ensure all engines(c) are running in the working directory
c[:].map(os.chdir, [FILE_DIR]*len(c))

# print list of engine ids
print(c.ids)

# restrict all the engines
bview = c.load_balanced_view()
register_parallel_backend('threading',
                        lambda : IPythonParallelBackend(view=
                        bview))

# loading the data
digits = load_digits()
# splitting the data to train and test
X_train, X_test, y_train, y_test = train_test_split(digits.data,
                                                    digits.target,
                                                    test_size=0.3)

# prepare the hyperparameter grid
param_grid = {
            "C": [c*(10**i) for i in range(1,14)],
            "gamma": [gamma*(10**i) for i in range(1,14)]
           }

# defining classifier with default hyperparameters
svc = SVC()
```

```
# defining GridSearchCV
search = GridSearchCV(svc,
                     param_grid,
                    return_train_score=True,
                    n_jobs=len(c))

# start timer
since = time.time()

# using parallel backend to start the parallel processing
with parallel_backend('threading'):
      search.fit(X_train, y_train)

# end the timer
time_taken = time.time() - since

# converting and saving the results to .csv file
print(f"Saving results to {FILE_DIR}")
results = search.cv_results_
results = pd.DataFrame(results)
results.to_csv(os.path.join(FILE_DIR,'scores.csv'))
print(f"Results Saved!")

# Display the time taken
print(f"Tuning Time: {time_taken}")
```

Let's examine step by step what's going on in the preceding Python script:

- Import all the libraries that were installed previously in our virtual environment.
- Since we need to use the profile name by which the ipython profile was created, we use argparser and take the name from the shell script as an argument.

- Initialize the client, giving it the profile name of the cluster to prepare all the engines. `c.ids` is used to get a list of all the engine IDs.
- Restrict load balancing across all the engines.
- Before running the algorithm parallelly, we define the back-end name as a string (`'threading'`), for which we use the function `register_parallel_backend`; later we'll use `'threading'` while running the training of the model.
- Load the dataset, split it to train and test, and define the grid for C and gamma.
- Initialize `SVC()` and define `GridSearchCV()`, with `n_jobs` set to either `-1` (use all available cores) or `len(total_engines)` (use a defined number of cores).
- We previously defined the name of the back end for parallel computation of jobs using `register_parallel_backend` with `'threading'`. We now use that name to run `parallel_backend` and Grid Search for 'C' and 'gamma' for SVM.
- Save the results.
- Execute the command `sbatch launch.sh python_script.py`; `sbatch` is a Slurm command that submits the written script to Slurm.
- Transfer the result files using the `scp` command.

And that is how we can optimize a huge number of hyperparameters on a cluster of computers while using it to its maximum potential. You can use random search in place of grid search as well. Using HPC will decrease your time consumption by a huge amount.

We reviewed some exhaustive and brute-force methods for hyperparameter optimization that would take a really long time if applied to problems like Neural Network Architecture Search and are not feasible. In later chapters we'll see some more algorithms and libraries that would be able to handle these complex tasks. However, in problems with fewer hyperparameters and a narrower search range, or where as a data scientist you can decide the approximate values or reduce the search space by looking at the dataset, these algorithms can be fruitful, and using them with machines such as those in HPC clusters can even increase their efficiency.

CHAPTER 3

Solving Time and Memory Constraints

We face two major problems while tuning hyperparameters:

- **Memory constraint**: Sometimes we have to deal with hundreds of gigabytes of data. We cannot store such a huge amount of data in RAM. While training a neural network, we send data in batches. One of the possible solutions is larger memory, which is not feasible always.
- **Time/computation constraint**: Let's say our data fits into memory, but we are training a deep neural network (DNN) or there is a huge search space for hyperparameter optimization. This can consume a great amount of time.

One solution to both of these problems is to use better hardware. For example, in the case of a deep neural network, graphics processing units (GPUs), tensor processing units (TPUs), and so forth can be used to accelerate the training. Although better hardware will solve the problem up to some extent, it's not always possible to work on such high-end machines. In Chapter 2, we distributed grid search over an HPC cluster, which solved the time constraint, but an HPC cluster is a costly resource. There are ways much easier, using different types of clusters.

T. Agrawal, *Hyperparameter Optimization in Machine Learning*,
https://doi.org/10.1007/978-1-4842-6579-6_3

In this chapter, we'll discuss easier alternatives to HPC that deal with both memory and time constraints. We'll mainly focus on distribution of training to efficiently use available resources.

We'll start with Dask, which is great for distributing machine learning processes and works best with scikit-learn. We'll then see easy ways to distribute neural networks using packages like PyTorch Distributed and Horovod.

Dask

Dask (`https://dask.org/`) is a library in Python for parallel computation. It uses *dynamic task scheduling*, similar to what we did using ipyparallel on the HPC cluster in Chapter 2, which addresses the problem of computational constraint. However, Dask is much more flexible than ipyparallel for distribution. Using dynamic task scheduling, we can distribute training over different machines (not just over multiple cores on the same machine).

But suppose we have a memory constraint because our data is so huge that it can't be loaded into memory at once. Dask solves this by offering *parallel collections* like Dask Dataframe, Dask array, and so forth, which distributes dataset into chunks that can either be used on a distributed environment or solve larger-than-memory problems. A process executed using Dask typically contains the following aspects:

Collection → Task Graph → Multi-Threading/ Processing or Distribution

First, a Dask collection is passed to a task graph, which is the complete pipeline of all the operations like preprocessing, hyperparameter optimization, evaluation, and so on. Here tasks can be parallelized or arranged in a serial manner. Finally, the scheduler can execute task graphs using dynamic task scheduling. If a single machine is used, multithreading/multiprocessing can be used to parallelize it over cores, and on a cluster, a task graph can be distributed over nodes. In the example of a task graph shown in Figure 3-1-1, operations (a) and (b) are independent, so both of them can be executed parallelly. Task 1, 2, and 3 inside (a) will be executed serially. If (a) is completed before (b), task 7 will wait for (b) to get completed.

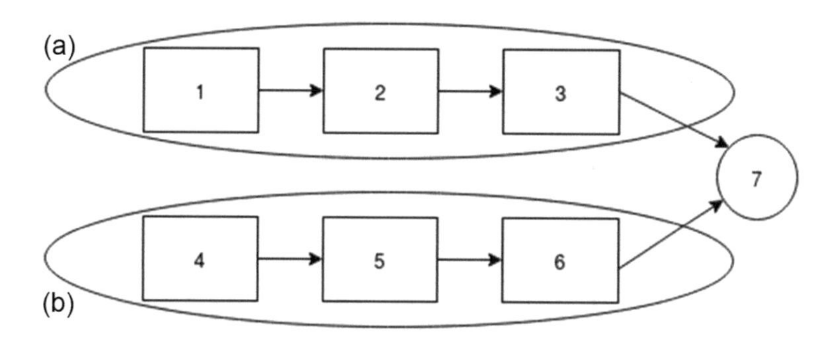

Figure 3-1-1. *An example of a task graph*

Dask Distributed

dask. distributed is a small library that extends to *dask* for dynamic task scheduling. While using dask we primarily use *Client*() from `dask.distributed`:

```
>>> from dask.distributed import Client
>>> client = Client()
```

Client() helps you connect to the distributed cluster. A Dask cluster is passed to *Client*(), depending upon the cluster type. Dask support several different cluster types, such as the following:

- *SSH*: If you have an unmanaged cluster, you need to connect to each machine using the SSH protocol. In that case, use the following:

```
>>> from dask.distributed import Client, SSHCluster
>>> cluster = SSHCluster(
                    ["localhost", "localhost", "localhost"],
                    connect_options={"known_hosts": None},
```

```
                        worker_options={"nthreads": 2},
                        scheduler_options={"port": 0,
                                          "dashboard_address":
                                          ":8797"},
                        )
>>> client = Client(cluster)
```

Here we define IPs for all the worker, `connect_options` can contain information like passwords for SSH connections.

- *Kubernetes*: Using a Kubernetes cluster is a quick and easy way to deploy distributed applications using Dask:

```
>>> from dask_kubernetes import KubeCluster
>>> cluster = KubeCluster.from_yaml('worker-template.yaml')
>>> cluster.scale(40) #add 40 worker nodes

>>> from dask.distributed import Client
>>> client = Client(cluster)
```

You can define worker machines as per your need, or even scale them as per workload using *cluster. adapt*() instead of *cluster. scale*().

Along with these cluster types, Dask also facilitates distribution over HPC, YARN (an Apache Hadoop cluster), and cloud-based clusters provided by Amazon, Google, and so forth.

By passing '`processes=False`' to *Client*(), a local cluster will be created and trials will be parallelized over cores:

```
>>> from dask.distributed import Client
>>> client = Client(processes=False)
>>> print(client)
Client
Scheduler: tcp://127.0.0.1:35053
Dashboard: http://127.0.0.1:8787/status
```

```
Cluster
Workers: 1
Cores: 4
Memory: 16.73 GB
```

This shows the client and cluster details. In this case, I have used a single machine as both scheduler and worker. On an actual cluster, you can utilize many more cores and workers, and much more memory.

A brilliant feature provided by Dask is the visualization of the distributed computing on the Dashboard (note the Dashboard address under the client info). Here you can visualize in real time the utilization of the cores once the search is executed. Later we'll take a look at some examples.

Parallel Collections

As you can see in Figure 3-1-2, chunks of a Dask dataframe consist of several small Pandas dataframes. Similarly, a Dask array consists of smaller NumPy arrays. You can decide the size of chunks such that they fit in memory.

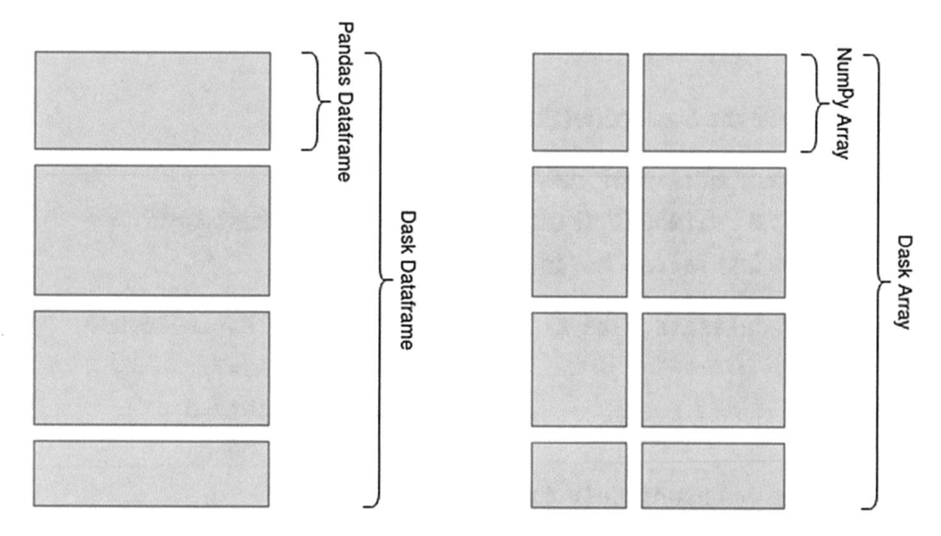

Figure 3-1-2. *Dask dataframe (left) and Dask array (right)*

Moreover, operations over Dask dataframe mimics API of pandas dataframe, and the same goes for arrays. For example:

```
#pandas
import pandas as pd
df = pd.read_csv("./dataset/train.csv")
print(df.Age.mean())

#dask
import dask.dataframe as dd
df = dd.read_csv("./dataset/train.csv")
print(df.Age.mean().compute())
```

The only difference is that you need to use `.compute()` to execute the operations in Dask. Not all NumPy and Pandas interfaces are supported though.

Since Daks parallel collections can help to solve memory constraints, let's model a large dataset that wouldn't otherwise fit in the memory, let alone train.

First initialize the client:

```
>>> from dask.distributed import Client
>>> client = Client(processes=False)
```

Now make the Dask collection:

```
>>> from dask_ml import datasets
>>> from dask_ml.model_selection import train_test_split
>>> import dask.array as da

>>> X, y = datasets.make_classification(n_samples=100000000,
                                        n_features=7,
                                        random_state=0,
                                        chunks=100000)
>>> classes = da.unique(y_train).compute()

>>> X_train, X_test, y_train, y_test = train_test_split(X, y)
```

Now we have a huge dataset, which has 100 million rows with seven features each and two classes. Figure 3-1-3 shows a data 5.6 GB large divided into 1000 chunks of 100,000 rows and 5.6 MB each.

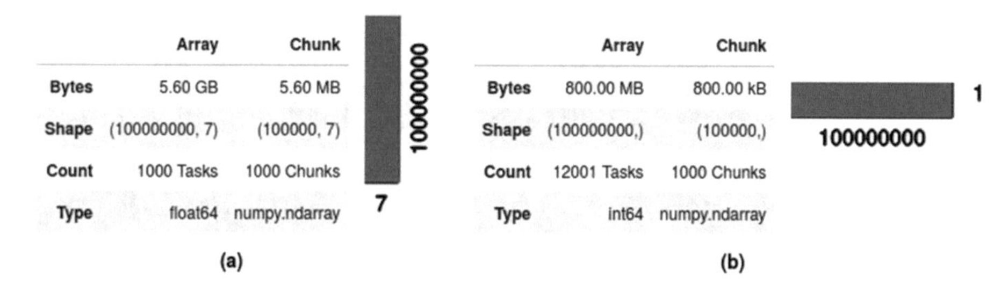

	Array	Chunk
Bytes	5.60 GB	5.60 MB
Shape	(100000000, 7)	(100000, 7)
Count	1000 Tasks	1000 Chunks
Type	float64	numpy.ndarray

	Array	Chunk
Bytes	800.00 MB	800.00 kB
Shape	(100000000,)	(100000,)
Count	12001 Tasks	1000 Chunks
Type	int64	numpy.ndarray

***Figure 3-1-3.** (a) is a representation of variable X and (b) is a representation of variable y*

The point of creating chunks is to not have to load the entire dataset in memory; only algorithms with 'partial_fit' in scikit-learn support this. I'll be using SGD Classifier to model the dataset:

```
>>> from sklearn.linear_model import SGDClassifier
>>> clf = SGDClassifier(loss='log', penalty="l2", tol=0.01)
```

Executing `clf.fit(X_train, y_train)` will iterate over the dataset a single time. To train the classifier further for multiple iterations, we can use a simple `for` loop. We'll have to wrap scikit-learn's SGD classifier into Dask's `Incremental` function, which manages the data so that the model will be trained in chunks.

```
>>> from dask_ml.wrappers import Incremental
>>> clf = Incremental(clf, scoring="accuracy")

>>> clf.fit(X_train, y_train, classes=classes)
```

Once you execute this, check out the Dask dashboard, where you can visualize various processes going on, as shown in Figure 3-1-4.

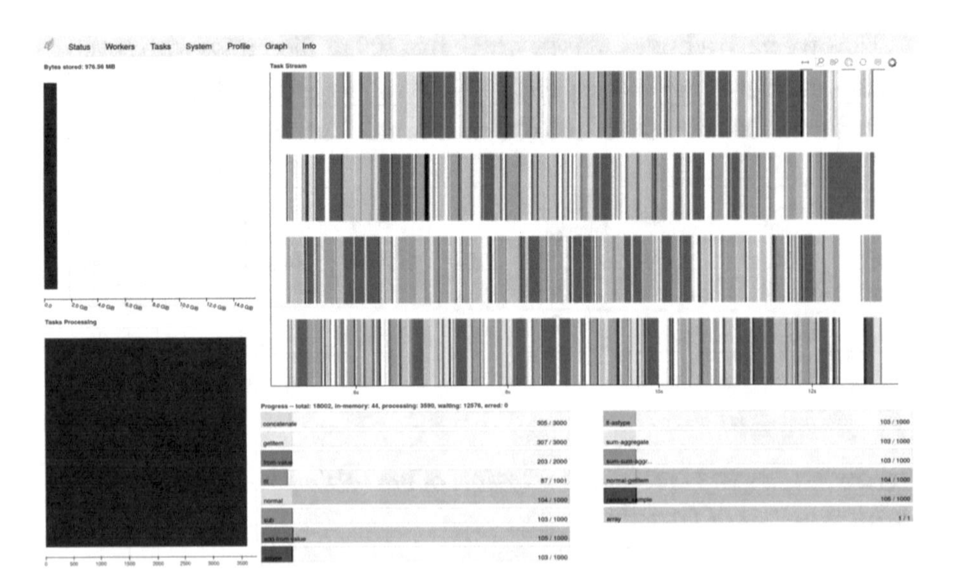

Figure 3-1-4. *The Dask dashboard Task Stream pane shows the four bars representing four cores.*

Dynamic Task Scheduling

Most of the algorithms implemented in scikit-learn are capable of using 'joblib', which provides thread/process-based distribution over the cores of a single machine. Using Dask, we can distribute these algorithms over a cluster, just like we did in the HPC cluster example in Chapter 2 using ipyparallel.

Dask, being much more flexible, provides support for parallelism on all different kinds of distributed systems, as depicted in 3-1-5.

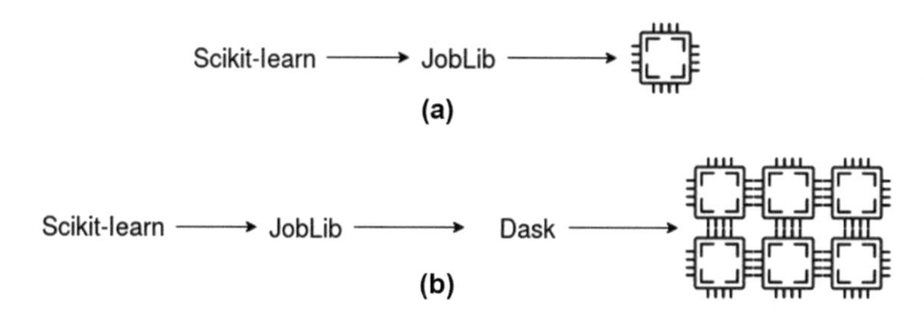

Figure 3-1-5. *(a) scikit-learn with the help of JobLib can distribute training over cores of a single machine. (b) With Dask, the same process can be distributed over cores of a single machine or even a cluster of machines*

Let's take an example of optimizing hyperparameters of support vector machine using both serial optimization using scikit-learn and distributing trails using Dask and compare the time:

```
from sklearn.datasets import load_digits
from sklearn.model_selection import GridSearchCV, train_test_
split
from sklearn.svm import SVC

X, y = load_digits().data, load_digits().target

X_train, X_test, y_train, y_test = train_test_split(X, y, test_
size = 0.3, shuffle=True)

c = 0.001
gamma = 1e-10
param_grid = {
          "C": [c*(10**i) for i in range(1,14)],
          "gamma": [gamma*(10**i) for i in range(1,14)]
         }

clf = SVC(kernel='rbf')
search = GridSearchCV(clf, param_grid, cv=3)
```

In the preceding code, we have loaded the 'digits' dataset, and we'll be optimizing 'C' and 'gamma' using `GridSearchCV()`. First, let's use just ScikitLearn example without Joblib.

```
>>> import time
>>> since = time.time()
>>> model = search.fit(X_train, y_train)
>>> print(time.time()-since)
```

The total number of trials would be 169 across the grid of hyperparameters, taking 75.01 seconds. Note that we are running Grid Search serially.

Now let's use Dask to parallelize our trials over the cores:

```
>>> import joblib
>>> import time
>>> from dask.distributed import Client
>>> client = Client(processes=False)

>>> since = time.time()
>>> with joblib.parallel_backend('dask', scatter=[X_train, y_train]):
            model = search.fit(X_train, y_train)
>>> print(time.time()-since)
```

Now JobLib will use Dask's client, which is, in this case, a local cluster for distribution of 169 possible combinations over the cores. The time taken for the same task was 34.73 seconds. This is a huge improvement with just four cores, and hence there's a huge scope of improvement on an actual cluster of machines. Check the Dask dashboard to visualize the core usage, as shown in Figure 3-1-6.

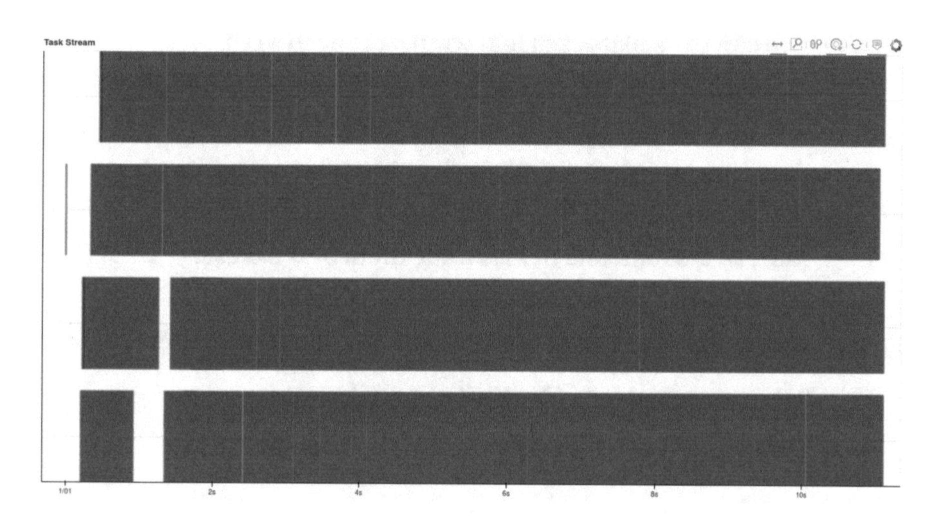

Figure 3-1-6. *Task Stream pane showing four horizontal bars denoting four cores*

Note All of the previous experiments are performed on a local machine, where both the scheduler and cluster are the same. In the example presented in the "Dynamic Task Scheduling" section, we could use scikit-learn alone to distribute over the cores of a single machine instead of using Dask.

Hyperparameter Optimization with Dask

As we saw in last few examples, we can use Dask for hyperparameter optimization to solve both time and memory constraints. Let's review the algorithms for hyperparameter optimization in scikit-learn and their distribution:

- We can use plain and simple scikit-learn's Random Search and Grid Search and pass the argument `'n_jobs'=-1` to use all cores on a single machine. This solves the compute/time constraint.

- We can use scikit-learn and wrap its code in `with parallel_backend('dask'):` as we did in the example in the "Dynamic Task Scheduling" section, be it Random Search or Grid Search or any other algorithm implemented with '*joblib*'. We can either use all cores on a single machine or distribute them over a cluster, depending on how dask.distributed's `client()` is defined. This reduces time even more if a cluster is used.

Now let's see what else Dask offers us for hyperparameter optimization.

Dask Random Search and Grid Search

We cannot use Random Search or Grid Search provided by scikit-learn to train a large dataset because they do not support 'partial_fit'. However, Dask gives us drop-in replacements for both of these algorithms, `dask_ml.model_selection.GridSearchCV` and `dask_ml.model_selection.RandomizedSearchCV`. Interface for respective algorithms in both Dask and Scikit-Learn is similar, but the one from Dask implements 'partial_fit', so that we can wrap a machine learning algorithm from scikit-learn in `Incremental` and pass it to these hyperparameter optimization algorithms.

The setting can not only train the model in chunks of data but also distribute it on a cluster, solving both time and memory issues. The steps are pretty much a straightforward script, the same as what we did before:

1. Define a client.
2. Define a search space.
3. Make a huge dataset to test our distributed model.
4. Train Test Split.
5. Define the ML algorithm which uses partial fit, like SGD classifier.

6. Wrap the ML algorithm in `Incremental` so that data can be managed and trained in chunks while distributing it to cores/cluster.

7. Use Grid Search on top of that which is imported from Dask since it implements 'partial_fit'.

8. Train the model under `joblib.parallel_backend` so that it can be further distributed.

However, there's an interesting problem I faced while following these steps. When I executed the code I checked the Dask dashboard. The memory started to fill up. The problem was due to the accuracy score. After I used accuracy metrics from 'dask_ml', the problem was solved.

Let's check out an example:

```
from dask_ml import datasets
from dask_ml.wrappers import Incremental
from dask_ml.model_selection import train_test_split,
GridSearchCV
from dask_ml.metrics import accuracy_score

from sklearn.metrics import make_scorer
from sklearn.linear_model import SGDClassifier

import joblib

import dask.array as da
from dask.distributed import Client
client = Client(processes=False)
print(client.dashboard_link)

param_grid = {
            "penalty": ['l1', 'l2'],
            "tol": [1e-2, 1e-3, 1e-4]
            }
```

```
X, y = datasets.make_classification(n_samples=100000000,
                                    n_features=7,
                                    random_state=0,
                                    chunks=100000)

# providing an accuracy metrics from 'dask_ml'
scorer = make_scorer(accuracy_score)

X_train, X_test, y_train, y_test = train_test_split(X, y)

clf = SGDClassifier(loss='log')
clf_wrap = Incremental(clf, scoring=scorer)
searh_clf = GridSearchCV(clf_wrap, param_grid, cv=3)

with joblib.parallel_backend('dask'):
      model = searh_clf.fit(X_train, y_train)
```

Note After some poking around, I found out that while score calculation by default 'SGDClassifier.score' was being used which was converting chunks of 'dask arrays' to 'ndarray', which was resulting in high memory usage. When I used a scorer from 'dask_ml', it solved the issue.

So, you need to take care that the chunks of your data are not converted to 'numpy array'; otherwise it would end up filling memory. The idea of using a Dask dataframe and Dask array here is to not fill memory.

Incremental Search

This is another really good approach to search hyperparameters. As we know, Dask can divide the data into chunks and the algorithms using 'partial_fit' can train small data at once. Incremental Search Algorithm uses this concept to its benefit. It trains several models on smaller chunks of datasets on a variety of sets of hyperparameters. It continues with further training on only the best-performing set of hyperparameters. However, there's a drawback to this method: what if the hyperparameter starts to perform better at a later stage? For example, in Figure 3-1-7 we have two hyperparameters, h1 and h2, and we train our model up to the ninth chunk of data. Initially h1 was performing better, but later h2 started performing better. But if we would have stopped the training at the second chuck, we wouldn't know this and would have discarded h2.

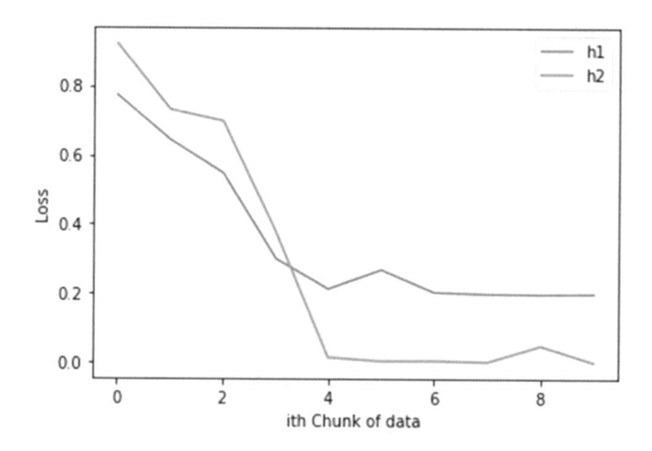

***Figure 3-1-7.** Loss decreasing as more chunks of data are passed for training*

Using incremental search is quite simple, as shown next. The interface is similar to that of GridSearchCV or RandomSearchCV.

```
from dask_ml.model_selection import IncrementalSearchCV
from sklearn.linear_model import SGDClassifier

param_grid = {
            "penalty": ['l1', 'l2'],
            "tol": [1e-1, 1e-2, 1e-3, 1e-4, 1e-5, 1e-6]
            }

clf = IncrementalSearchCV(SGDClassifier(), param_grid)

#fit the data to train
```

Figure 3-1-8 shows a trial vs accuracy plot comparing Random Search and Incremental Search.

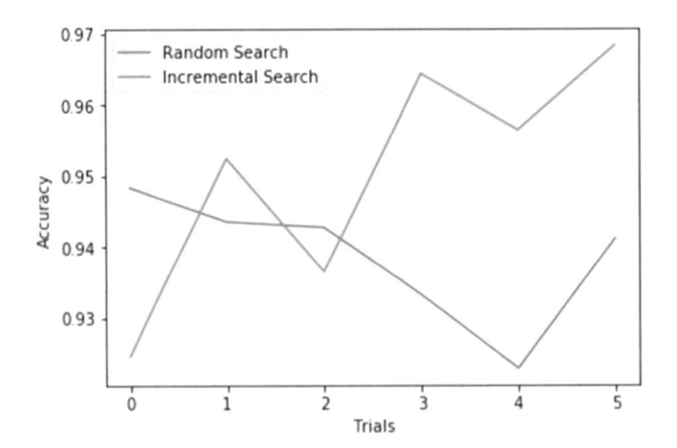

***Figure 3-1-8.** A comparison between Random Search and Incremental Search. Digits dataset is used from sklearn, trained on SGD classifier. Results are compared on the first five trials*

Successive Halving Search

Success halving search is somewhat similar to incremental search. A time budget (B) is uniformly assigned to all sets of hyperparameters. We start training the model on a limited amount of data, and we start by using all sets of hyperparameters. Once all the models are trained, half of the worst-performing sets are discarded. In the next iteration, the remaining half are now trained on more data than before, evaluated and again half of the previous half is discarded.

This process goes on until one best set remains, thereby allocating more resources to better-performing hyperparameters. Use Dask implementation of Successive Halving as follows:

```
from dask_ml.model_selection import SuccessiveHalvingSearchCV
from sklearn.linear_model import SGDClassifier
```

```
clf = SuccessiveHalvingSearchCV(SGDClassifier(), param_grid,
n_initial_iter=2)
#call fit to train data
```

Here, we have to pass `n_initial_iter` to our search function, which defines the number of times 'partial_fit' is called initially.

The same problem can persist as in incremental search, though, where we might pick a wrong set of hyperparameters if we stop early.

Hyperband Search

Hyperband[1] is a bandit-based approach for solving the problem of hyperparameter optimization. The bandit-based approach addresses our problem perfectly here: we have a limited amount of resources but we need to allocate them to all our trials efficiently. Again we spend more time on better-performing models instead of wasting our resources and time on poor configuration of hyperparameters.

Hyperband is an extended version of successive halving. In successive halving, we have a fixed budget (B) for our sets of hyperparameters (n), and 'B/n' resources are allocated uniformly to all the sets in 'n'. But how we should choose 'n' remains a problem. There are two possibilities:

- We should consider a lesser value of 'n' so that the resources provided to each configuration are more and the training time is longer. But then less search space would be covered. This case can be favorable when the accuracy score is more dependent on the training data than on hyperparameters.

[1]"Hyperband: A Novel Bandit-Based Approach to Hyperparameter Optimization," L. Li, K. Jamieson, G. DeSalvo, A. Rostamizadeh, and A. Talwalkar, *Journal of Machine Learning Research* 18 (2018) 1-52.

- We should consider a higher value of 'n' so that more search space is covered with less amount of training. A high value of 'n' would be better if a change in hyperparameters results in a huge change in accuracy/ loss scores, unlike the case in Figure 3-1-7, where at a later stage we needed to train longer for the loss score to get saturated.

Hyperband addresses this problem of "n v/s B/n" by keeping the value of B a constant and changing the value of 'n'. A larger value of 'n' thus results in aggressive early stopping, because the budget per set of hyperparameters is reduced. In Hyperband we change the value of 'n' each time we iterate for successive halving. There's a nested loop; while the inner loop performs successive search, the outer loop iterates over different values of 'n'. There are lots of possibilities for parallelization, which is exploited in the Dask implementation of Hyperband:

```
from dask_ml.model_selection import HyperbandSearchCV
from sklearn.linear_model import SGDClassifier

clf = HyperbandSearchCV(SGDClassifier(), param_grid)
#call fit to train data
```

Two important arguments given to Hyperband are `max_iter`, which defines the number of times 'partial_fit' is called, and the chunk size of 'partial_fit'. Both of these arguments are determined by rule of thumb, where `max_iter` is equal to the number of hyperparameter combinations (n_param) and chuck size is n_param/n_examples, where n_examples is the number of samples the model is trained on; for example, if X_train is trained in five iterations, n_examples = 5*len(X_train).

Note When using these optimization algorithms (Incremental search, Successive Halving, Incremental Search) you can only use modeling algorithms that use 'partial_fit', because the whole idea is to stop training early, which we cannot do with other algorithms. Although you can use RandomSearchCV and GridSearchCV to distribute trials over a cluster, you can't use them when data is trained in chunks (larger than memory case); in that case you need 'partial_fit' again. These methods can be really helpful while working with neural networks, since they are trained on batches.

Dask serializes several objects in order to distribute. scikit-learn and PyTorch would work better with these algorithms, since they work on the pickle protocol, unlike TensorFlow and Keras. But when using Keras and TensorFlow, you can apply other algorithms like GridSearchCV, RandomSearchCV, and more. This we are going to discuss in the next section.

Distributing Deep Learning Models

Deep learning models are quite costly to train. Distributing neural network training over a cluster and later applying hyperparameter optimization on top of that can help us save a lot of time. Deep learning frameworks have modules for distribution; for example, TensorFlow Distributed (which extends to TensorFlow and Keras), PyTorch Distributed (which extends to PyTorch), Horovod, and so forth. Distributing with TensorFlow Distributed is a pain. You have to create parameter servers and change a lot of code. In this section we'll discuss PyTorch Distributed and Horovod to distribute deep neural networks while training the MNIST dataset. If you want a quick refresher on using the PyTorch API, refer to Appendix II.

PyTorch Distributed

The main difference between distributing on a machine and a cluster is that in a cluster we need to have a back-end communication API to communicate between nodes. This is one of the strong aspects of PyTorch. PyTorch supports all three major back-end communication APIs: NCCL (NVIDIA Collective Communications Library, pronounced "Nickel"), Gloo, and MPI (Message Passing Interface). The following example is strongly based on a GitHub repo (`https://github.com/seba-1511/dist_tuto.pth`):

```
import torch
import torch.nn as nn
import torch.nn.functional as F
import torch.optim as optim
import torch.distributed as dist
from torch.autograd import Variable
from torch.multiprocessing import Process

import os
from math import ceil
from random import Random
from torchvision import datasets, transforms

def init_processes(rank, size, fn, backend="gloo"):
      os.environ['MASTER_ADDR'] = '127.0.0.1'
      os.environ['MASTER_PORT'] = '29500'
      dist.init_process_group(backend, rank=rank,
      world_size=size)
      fn(rank, size)
```

We start by defining the communication back end; we'll be using 'gloo'. Next we define the master address and master port so that all nodes report to one central master node. Each time that function `init_processes` is executed, a new process group is created and function `fn()` is executed. In our case `fn` would be a function to train the neural network, so we would later write a function that would train the model on subsamples and later average the gradients.

Note PyTorch documentation suggests as a rule of thumb to use 'gloo' when using CPUs and 'nccl' while training on GPUs.

```
class Net(nn.Module):

    def __init__(self):
        super(Net, self).__init__()
        self.conv1 = nn.Conv2d(1, 10, kernel_size=5)
        self.conv2 = nn.Conv2d(10, 20, kernel_size=5)
        self.conv2_drop = nn.Dropout2d()
        self.fc1 = nn.Linear(320, 50)
        self.fc2 = nn.Linear(50, 10)

    def forward(self, x):
        x = F.relu(F.max_pool2d(self.conv1(x), 2))
        x = F.relu(F.max_pool2d(self.conv2_drop(self.
        conv2(x)), 2))
        x = x.view(-1, 320)
        x = F.relu(self.fc1(x))
        x = F.dropout(x, training=self.training)
        x = self.fc2(x)
        return F.log_softmax(x)
```

We need to create a neural network to build the model, so here we define a simple convolutional network:

```
class Partition(object):

    def __init__(self, data, index):
        self.data = data
        self.index = index

    def __len__(self):
        return len(self.index)

    def __getitem__(self, index):
        data_idx = self.index[index]
        return self.data[data_idx]

class DataPartitioner(object):

    def __init__(self, data, sizes=[0.7, 0.2, 0.1], seed=1234):
        self.data = data
        self.partitions = []
        rng = Random()
        rng.seed(seed)
        data_len = len(data)
        indexes = [x for x in range(0, data_len)]
        rng.shuffle(indexes)

        for frac in sizes:
            part_len = int(frac * data_len)
            self.partitions.append(indexes[0:part_len])
            indexes = indexes[part_len:]

    def use(self, partition):
        return Partition(self.data, self.
        partitions[partition])
```

Our model is being trained parallelly, so we need to update all of the gradients once it's trained on a batch of data. So once training on a batch is done, we collect all gradients and take their average. Data is sent to all nodes in equal fractions so that the convergence by each model remains uniform. The partitioning is done on the basis of the number of processes running. We can use the classes identified in the previous listing to get a certain fraction of data.

```
def partition_dataset():
      transformations = [transforms.ToTensor(),
                         transforms.Normalize((0.1307, ),
                         (0.3081, ))
                        ]
      dataset = datasets.MNIST('./data',
                               train=True,
                               download=True,
                               transform=transforms.
                               Compose(transformations)
                              )
      size = dist.get_world_size()
      bsz = int(8 / float(size))
      partition_sizes = [1.0 / size for _ in range(size)]
      partition = DataPartitioner(dataset, partition_sizes)
      partition = partition.use(dist.get_rank())
      train_set = torch.utils.data.DataLoader(
                                   partition, batch_size=bsz,
                                   shuffle=True)
      return train_set, bsz
```

This function loads the MNIST dataset and uses `DataPartitioner()` to partition the data. `dist.get_world_size()` returns the number of processes in the current process group. So if three processes are running, we would have three partition sizes of fraction 0.33 each. Similarly for batch size, we divide the required batch size by the number of processes:

```
def average_gradients(model):
    size = float(dist.get_world_size())
    for param in model.parameters():
        dist.all_reduce(param.grad.data, op=dist.reduce_
        op.SUM)
        param.grad.data /= size
```

Here we average the gradients. all_reduce updates parameters in all the distributed models. And Now we define the function which will optimize the model, and we'll pass to init_processes().

```
def run(rank, size):
    torch.manual_seed(1234)
    train_set, bsz = partition_dataset()
    model = Net()
    model = model
    optimizer = optim.SGD(model.parameters(), lr=0.01,
    momentum=0.5)

    num_batches = ceil(len(train_set.dataset) / float(bsz))
    for epoch in range(10):
        epoch_loss = 0.0
        for data, target in train_set:
            data, target = Variable(data),
            Variable(target)
            optimizer.zero_grad()
            output = model(data)
            loss = F.nll_loss(output, target)
            epoch_loss += loss.item()
            loss.backward()
            average_gradients(model)
            optimizer.step()
```

```
            print(f'Rank: {dist.get_rank()}, \
                    Epoch: {epoch}, \
                    Loss: {epoch_loss / num_batches}')
```

Each rank signifies the models that are being trained parallelly. We call `partition_dataset()` to get the train set. While iterating, we are using the function `average_gradients()` to average the gradients. Now, finally, we can start the distributed training by passing the function `run()` to `init_processes()`:

```
size = 3
processes = []
for rank in range(size):
      p = Process(target=init_processes, args=(rank, size, run))
      p.start()
      processes.append(p)

for p in processes:
      p.join()
```

This will start your distributed training. Using this, with a few edits according to networks and use cases, we can train deep learning models on a computer cluster.

Horovod

Horovod is an open source distributed deep learning training library that works with both PyTorch and TensorFlow/Keras. If you have code for undistributed training, you can use Horovod simply by adding only a few lines of code to make it distributed. (Refer to the documentation on how to distribute on different clusters.) Similar to PyTorch Distributed, Horovod is capable of using both Gloo and MPI and other back-end communications. In this section we'll examine what changes we need to make and how they are similar to the PyTorch implementation in the previous section.

Write a simple code for training a neural network in a simple machine in PyTorch. For this you can refer to the PyTorch section of Appendix II. Create the following functions:

- `Network()`: A class extended from `torch.nn.Module` to create a neural network architecture
- `train_epoch()`: A function that can train the neural network for a single machine

Also define these global variables: `batch_size`, `learning_rate`, `momentum`, and epochs. The following piece of code shows how we can distribute model training using horovod:

```
import horovod.torch as hvd
from sparkdl import HorovodRunner
from torch.utils.data.distributed import DistributedSampler

def train_hvd():
      hvd.init()
      device = torch.device('cuda' if torch.cuda.is_available()
                             else 'cpu')

      if device.type == 'cuda':
           torch.cuda.set_device(hvd.local_rank())

      transformation = [transforms.ToTensor(),
                        transforms.Normalize((0.1307,),
                        (0.3081,))]
      train_dataset = datasets.MNIST(
                              root=f'data-{hvd.rank()}',
                              train=True,
                              download=True,
                              transform=transforms.
                              Compose(transformation)
                              )
```

```
train_sampler = DistributedSampler(train_dataset,
                                   num_replicas=hvd.size(),
                                   rank=hvd.rank())
train_loader = torch.utils.data.DataLoader(train_dataset,
                          batch_size=int(batch_size/hvd.
                          size()),
                          sampler=train_sampler)
model = Network().to(device)
optimizer = optim.SGD(model.parameters(),
                      lr=learning_rate,
                      momentum=momentum)
optimizer = hvd.DistributedOptimizer(optimizer,
                    named_parameters=model.named_
                    parameters())

hvd.broadcast_parameters(model.state_dict(), root_rank=0)

for epoch in range(1, epochs + 1):
      train_epoch(train_loader)
```

The function `train_hvd()` is pretty much similar to what we did while using PyTorch Distributed. `hvd.rank()` gives the worker ID, so we create a separate root folder for every worker (rank) in our cluster.

Similar to the previous section in which we used `partition_dataset()` to create and divide data among workers, here we use `DistributedSampler()` to create partitions of the dataset as per the number of workers and give the object to Dataloader() to generate the dataset. Again, effective batch size is scaled based on number of workers.

After defining `optimizer`, we wrap it in `hvd.DistributedOptimizer()`, which is similar to `all_reduce()` that we used earlier. The gradients are averaged across models running on different nodes. `hvc.broadcast_parameters()` updates all the gradients with new gradients. It makes sure models on all ranks start with the same parameters.

Now to distribute this training across clusters, we'll use a simple interface provided by HorovodRunner:

```
hr = HorovodRunner(np=2)
hr.run(train_hvd)
```

Here, np defines the number of workers and hr.run() starts the distributed training.

These distribution methodologies can come in handy when you are working on optimizing hyperparameters in hundreds to thousands of dimensions. Time efficiency and resource utilization are both important aspects while optimizing hyperparameters.

In the next chapter, you'll see more advanced Bayesian-based optimization methods, which actually learn from their previous trials.

CHAPTER 4

Bayesian Optimization

In Chapters 2 and 3 we explored several hyperparameter tuning methods. Grid search and random search were quite straightforward, and we discussed how to distribute them to save memory and time. We also delved into some more-complex algorithms, such as HyperBand. But none of the algorithms that we reviewed learned from their previous history. Suppose an algorithm could keep a log of all the previous observations and learn from them. For example, suppose it could observe that our model is being optimized near certain values of hyperparameters and could exploit this valuable information and proceed to the hyperparameters nearest to those good-performing ones, hence learning from its history. By doing so, the algorithm would not waste time on bad-performing hyperparameters while reaching the best-performing hyperparameters. In this chapter we'll explore algorithms that have that capability.

Let's start with an example. Figure 4-1-1 shows a plot between two hyperparameters. Compare that to Figure 2-2-1 in Chapter 2, which represents a grid search in which we were going through a grid of selected parameters and, in a random search, randomly hitting a set of hyperparameters.

T. Agrawal, *Hyperparameter Optimization in Machine Learning*,
https://doi.org/10.1007/978-1-4842-6579-6_4

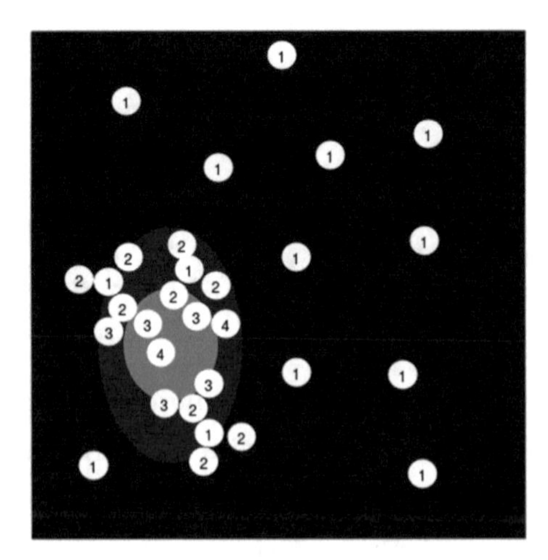

***Figure 4-1-1.** Plot between two hyperparameters, where darker area represents lesser accuracy and lighter area represents greater accuracy*

In Figure 4-1-1, the circles with number 1 are the hyperparameters chosen at random. Some of the 1s lying in the middle region between the darkest and lightest regions were observed to generate better models. So, instead of wasting our time on the rest of the 1 trials, we look at surrounding regions of better-performing 1 trials and train the model with 2s, and so on. In the end, we reach 4, where one of the 4s lies in the middle region and one lies almost at the center of the lightest region. Hence, we select the 4 trial lying in the lightest region as our best set of hyperparameters. Intuitively, this process seems better than the exhaustive methods you learned before.

Sequential Model-Based Global Optimization

It is notable that in machine learning, functions are expensive and slow to train and evaluate. In this section we'll look at sequential model-based global optimization (SMBO) to solve the problem of hyperparameter optimization. SMBO uses the approach of Bayesian optimization, which is used to keep track of previous evaluations and select the subsequent set of hyperparameters based on a probabilistic model.

$$p(y|x)$$

Our objective function (machine learning algorithm) being *f*, *y* is the score calculated by evaluating *f* on the set of hyperparameters *x*. In Bayesian optimization, essentially there are four important aspects (defined after the following step list): search space, objective function, probabilistic regression model, and acquisition function.

Whole Bayesian optimization can be summarized in the following steps:

1. Build a regression model.
2. Initialize some random sets of hyperparameters (in the case of the first trial, because we need to feed initial hyperparameters from somewhere).
3. Evaluate the model on the set of hyperparameters suggested by the acquisition function (if the first trial, choose hyperparameters from step 2) and calculate the score on the objective function.
4. Update the surrogate model as per the new suggested hyperparameters and scores.
5. Repeat steps 3 and 4, for a defined number of iterations.

A *search space* (*X*), as you learned in Chapter 1, is a defined range where we provide hyperparameter optimization algorithms a range to choose. Depending on the hyperparameter, ranges can be continuous or discrete. For example, choosing a kernel function in SVM is a discrete hyperparameter, but gamma is chosen from a continuous distribution. Search spaces can be really complicated. For example, choosing the number of nodes in each hidden layer in a neural network depends on the number of hidden layers.

An *objective function* (*f*) is a function that trains a machine learning model on a given set of hyperparameters and the output is either an accuracy score or a loss score depending on the acquisition function. In the following example, we are calculating the accuracy score, if the returned value is minimized, we'll maximize the accuracy score by minimizing its negative.

```
X, y = load_data()  # X and y are some preprocessed data
X_train, X_test, y_train, y_test = train_test_split(X, y, test_
size = 0.3)

def objective_function(hyperparameters):
      C = hyperparameters["C"]
      gamma = hyperparameters["gamma"]

      model = SVC(C=C, gamma=gamma)
      model.fit(X_train, y_train)

      score = model.score(X_test, y_test)
      # since we want to maximize score, taking it's negative
      return -score
```

Probabilistic Regression Model ($p(y|x)$ *or* M) also called Surrogate function is built using previous evaluations and is a probabilistic modeling of the objective function, so each iteration updates the surrogate by evaluating hyperparameters on the objective function. A surrogate function is *less costly to evaluate* in comparison with an objective function, and a surrogate function decides the next set of hyperparameters to be evaluated by the objective function, thus reducing the cost of optimization. A surrogate can be formulated by different methods, such as Gaussian process (GP), Random forest, or tree-structured Parzen estimator (TPE). Here's a brief overview of how these surrogates are formed:

- In Gaussian process, function *f* is assumed to be a realization of Gaussian distribution, where predictions follow a normal distribution. GP models $p(y|x)$ directly.
- In TPE, $p(y|x)$ is modeled on both $p(x|y)$ and $p(y)$. We'll discuss TPE in more detail later in this chapter.

An *acquisition function* (*S*) selects the next set of hyperparameters using the surrogate model and the predicted loss score on the previous set of hyperparameters. There are several acquisition functions, such as probability of improvement, expected improvement, conditional entropy of minimizer, and bandit-based criteria. The most commonly used is expected improvement:

$$EI_{y*}(x) := \int_{-\infty}^{y*} max\left(y^* - y, 0\right).p\left(y|x\right).dy$$

Here, *EI* is being modeled by surrogate and the loss score. y^* is some threshold value, while $y = f(x)$ is the score obtained from the objective function on the proposed set of hyperparameters *x*. $p(y|x)$ is the surrogate model. A positive value of the preceding integral means that chances are good that the proposed hyperparameters would yield better results. On the proposed set *x*, if *y* increases negatively, EI will be positive, hence indicating a better choice of hyperparameters.

Here is a generalized pseudo-code template of how the SMBO method works:

Input: $f, \mathcal{X}, S, \mathcal{M}$
$\mathcal{D} \leftarrow$ INITSAMPLES$(f, \mathcal{X})$
for $i \leftarrow |\mathcal{D}|$ **to** T **do**
  $p(y \mid \mathbf{x}, \mathcal{D}) \leftarrow$ FITMODEL$(\mathcal{M}, \mathcal{D})$
  $\mathbf{x}_i \leftarrow \arg\max_{\mathbf{x} \in \mathcal{X}} S(\mathbf{x}, p(y \mid \mathbf{x}, \mathcal{D}))$
  $y_i \leftarrow f(\mathbf{x}_i)$ ▷ Expensive step
  $\mathcal{D} \leftarrow \mathcal{D} \cup (\mathbf{x}_i, y_i)$
end for

In this pseudo-code, *f* is the objective function, *X* is the search space for hyperparameters, *S* is the acquisition function, and *M* is the regression model (surrogate).

First we initialize *D* with some random samples from the search space; *D* store history of evaluations in the form of (x_i, y_i), where x_i represents the subsequent sets of hyperparameters and y_i represents the loss scores.

We now run the loop for defined number of trials *T*. First the surrogate is updated using history *D*. Now *S* suggests a set of hyperparameters *xi*. *xi* is sent to *f* and a loss score is calculated. History is now saved in *D*, which would be used again to update the surrogate.

After *T* trials, we would have the best set of hyperparameters.

You don't have to implement these methods, because all the probabilistic regression models can be found implemented in different libraries. For example, Hyperopt[1] implements a TPE, Spearmint[2] and MOE[3] implement a Gaussian process, and SMAC[4] implements a random forest-based surrogate.

Next we'll discuss in detail the working of the tree-structured Parzen estimator along with the expected improvement acquisition function.

Tree-Structured Parzen Estimator

Tree-structured Parzen estimator is a popular Bayesian optimization approach that uses the expected improvement acquisition function[5]. In TPE, $p(y|x)$ is modeled over $p(x|y)$ and $p(y)$ following Bayes' theorem, unlike GP where $p(y|x)$ is directly modeled

[1]`https://github.com/hyperopt/hyperopt`

[2]`https://github.com/JasperSnoek/spearmint`

[3]`https://github.com/Yelp/MOE`

[4]`https://github.com/automl/SMAC3`

[5]`https://papers.nips.cc/paper/4443-algorithms-for-hyper-parameter-optimization.pdf`

$$p(y|x) = \frac{(p(x|y) * p(y))}{p(x)} \quad \text{(Equation 4.2.1)}$$

$p(x| y)$ is represented as

$$p(x| y) = l(x) \text{ if } y < y^* \quad \text{(Equation 4.2.2)}$$

and

$$p(x| y) = g(x) \text{ if } y > = y^* \quad \text{(Equation 4.2.3)}$$

Here, y^* is a quantile that is called *threshold loss score*. $l(x)$ and $g(x)$ are hyperparameter distributions. Hence, on certain sets of hyperparameters (x_i), if predicted loss score (y) is less than y^*, it means those sets lie in distribution $l(x)$, and if y is greater than y^* then they lie in distribution $g(x)$. We can understand this clearly by Figure 4-2-1.

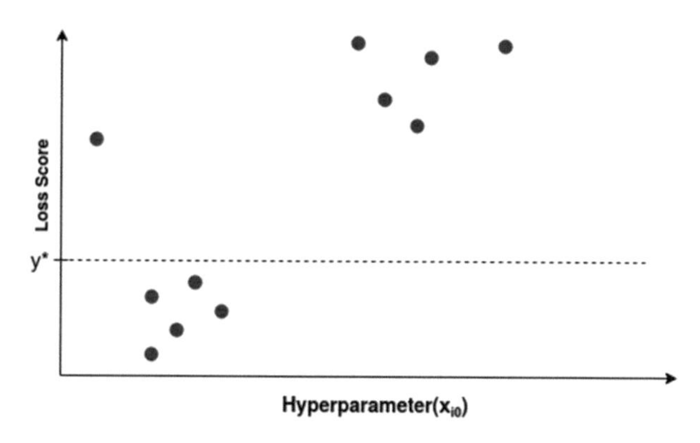

Figure 4-2-1. *On hyperparameter x_{i0} points below y^* are from distribution $l(x)$ and points above y^* are from distribution $g(x)$*

The value of y^* is chosen such that it is larger than the best observed value of $f(x)$, $y < y^*$, and therefore we have to find sets of hyperparameters that lie in the $l(x)$ distribution so that the predicted loss score is less than y^*.

Let

$$p(y < y^*) = \gamma \quad \text{(Equation 4.2.4)}$$

and continuous marginal probability be

$$p(x) = \int_R p(x|y).p(y).dy \quad \text{(Equation 4.2.5)}$$

Now we optimize EI:

$$EI_{y*}(x) = \int_{-\infty}^{y^*} (y^* - y).p(y|x).dy = \int_{-\infty}^{y^*} (y^* - y)\frac{p(x|y).p(y)}{p(x)}.dy$$

(Equation 4.2.6)

Equation 4.2.5 can be written as follows by using Equations 4.2.2, 4.2.3, and 4.2.4:

$$p(x) = \int_R p(x|y).p(y).dy = \gamma.l(x) + (1-\gamma).g(x) \quad \text{(Equation 4.2.7)}$$

Now that denominator term in Eq. 4.2.6 is simplified to Equation 4.2.7, which doesn't depend on *y*. Let's simplify the numerator:

$$\int_{-\infty}^{y^*} (y^* - y).p(x|y).p(y).dy = l(x)\int_{-\infty}^{y^*} (y^* - y).$$
$$p(y).dy = \gamma.y^*.l(x) - l(x)\int_{-\infty}^{y^*} p(y).dy \quad \text{(Equation 4.2.8)}$$

Putting together Equations 4.2.7 and 4.2.8, we get the value of expected improvement:

$$EI_{y*}(x) = \frac{\gamma.y^*.l(x) - l(x)\int_{-\infty}^{y^*} p(y).dy}{\gamma.l(x) + (1-\gamma).g(x)}$$

Simplify it further and we get:

$$EI_{y*}(x) = \frac{l(x)\left[\gamma . y^* - \int_{-\infty}^{y^*} p(y).dy\right]}{\gamma . l(x) + (1-\gamma).g(x)}$$

Therefore:

$$EI_{y*}(x) \propto \left[\gamma + \frac{g(x)}{l(x)}.(1-\gamma)\right]^{-1} \quad \text{(Equation 4.2.9)}$$

Equation 4.2.9 significantly tells us that expected improvement is inversely proportional to the ration $g(x)/l(x)$. This means we would prefer our hyperparameters (x) to lie in distribution $l(x)$ instead of $g(x)$ to increase the EI.

And thus on each iteration, candidates are drawn from $l(x)$ and return the one with least loss score, hence selecting best hyperparameters.

Now that you have a basic idea of how in the Bayesian method we model surrogate functions and acquisition functions, we'll next look at an open source library that implements TPE, Hyperopt.

Hyperopt

Hyperopt is a brilliant open source library for distributed asynchronous hyperparameter optimization that implements algorithms like random search, TPE, and adaptive TPE. In this section we'll focus on how we can use the Hyperopt library to optimize hyperparameters.

Hyperopt handles awkward search spaces, which includes searching over both discrete and continuous values. We can use the library to search between algorithms and find the best set of hyperparameters for those algorithms. When working on problems in deep learning, we deal in hundreds of dimensions, and to exploit the full potential of deep networks, we need the hyperparameter setting to be optimal. Using grid search or random search would not be an option, because each training of the network is quite costly. In such cases, using Bayesian optimization can be the best option.

To use Hyperopt we need a search space and an objective function. Let's take a simple example:

$$f(a,b) = a^2 - b^2$$

Here, we minimize $f(a, b)$ such that $a \in [-2, 3]$ and $b \in [-1, 2]$.

Let's use Hyperopt to optimize this problem:

```
from hyperopt import tpe, fmin, hp


def objective_func(args):
        a = args['a']
        b = args['b']
        f = a**2 - b**2
        return f

range_a = hp.uniform('a', -2, 3)
range_b = hp.uniform('b', -1, 2)

space = {'a': range_a,
         'b': range_b}

best = fmin(objective_func, space, algo=tpe.suggest,
max_evals=100)
```

The preceding piece of code uses TPE to find the best values of a and b such that $f(a, b)$ is minimum. We can see that function would be minimum at $a = 0$ and $b = 2$, and the minimum value would be $f_{min} = -4$. Let's see how Hyperopt in 1000 trials approached the problem.

Figure 4-3-1 shows that the values of a and b are saturating over 0 and 2 respectively, and the value of f is saturating over −4 in the early trials. This was an easy function where optimal values were integers. For more complicated functions where values must be picked from a continuous distribution, TPE proves to be efficient.

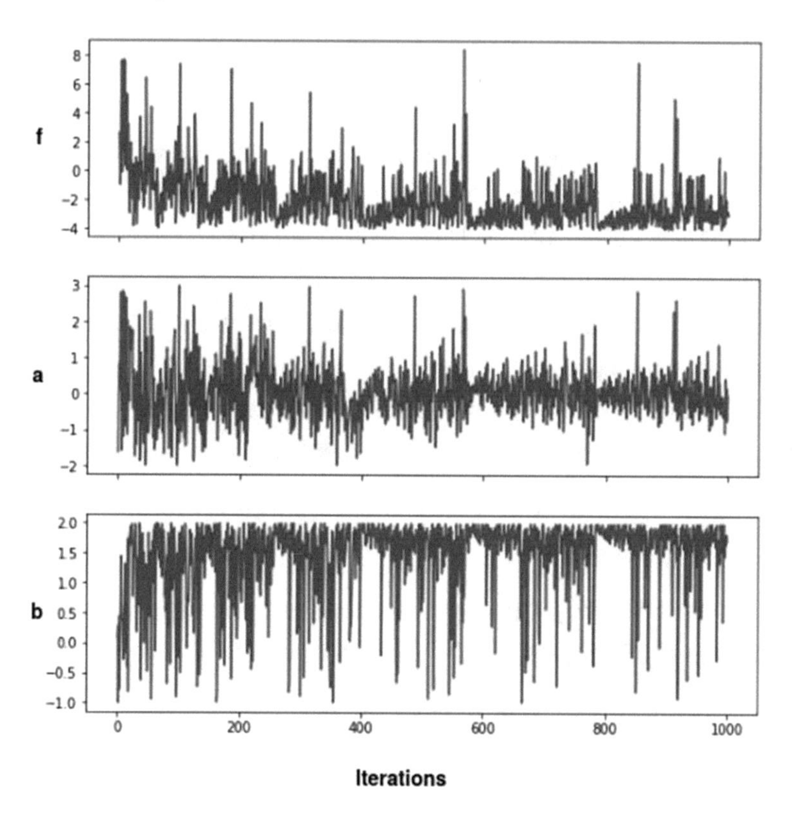

Figure 4-3-1. *Plotting value of f, a and b over 1000 iterations using TPE*

As mentioned earlier, to use Hyperopt we must define an objective function and search space. In the previous code, our objective function takes hyperparameters as inputs and outputs a score that we want to minimize. To create the search space, for each hyperparameter we must use distribution in the form of a Hyperopt object. We have a wide variety of options, uniform, normal, loguniform, lognormal, and so on, which we'll discuss in more detail in the next section. Finally, we pass both the objective function and search space to the `fmin()` function while using the algorithm TPE for optimization. We also decide the number of trials, just as we did when using random search. A dictionary is returned that give the best trials out of all the iterations.

Search Space

fmin() passes only one parameter to the objective function, so we need to stuff all the hyperparameter ranges in either one of the dictionaries, list or tuples. Unlike scikit-learn's Grid Search and Random Search, `fmin()` does not support just any iterable distribution; all the hyperparameters should be objects of Hyperopt's *hp* module. Follow Figure 4-3-2 for visualization of different distributions.

Following are some of the functions that give a certain value from different types of distributions, for our hyperparameter searches:

- `hp.choice()`: chooses one of the options from the given list
- `hp.randint()`: Chooses a random integer out of a range of integers
- `hp.uniform()`: Returns a value between a range, the distribution is uniform between two given numbers
- `hp.loguniform()`: Returns a value such that its logarithm is uniformly distributed between two given numbers
- `hp.normal()`: Returns a value from a Gaussian distribution as per mean and standard deviation
- `hp.lognormal()`: logarithm of the returned value is normally distributed as per mean and standard deviation

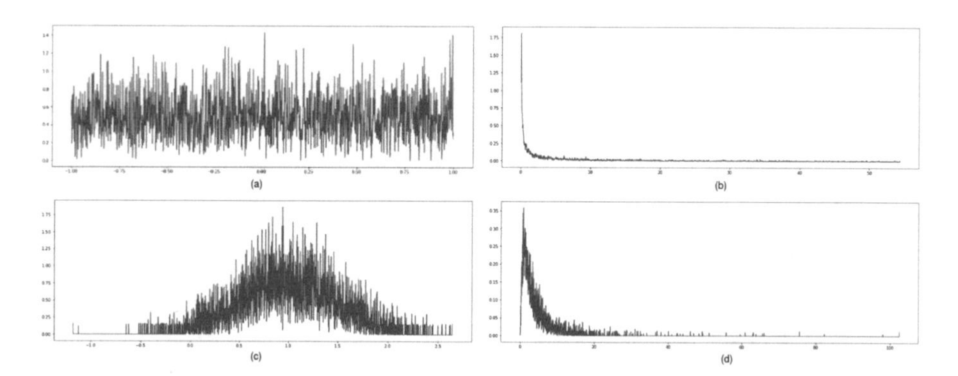

Figure 4-3-2. *These graphs are plotted using Hyperopt distributions. Graph (a) shows a uniform distribution between -1, 1. Graph (b) shows a loguniform distribution between -3, 4. Graph (c) shows a normal distribution with mean 1 and standard deviation 0.5. Graph (d) shows a lognormal distribution with mean and standard deviation both 1*

Tip *loguniform*(*label*, *a*, *b*) can be written as *exp*(*uniform*(*label*, *a*, *b*)), where *a* and *b* are the lower and upper limit, respectively. Distribution is between e^a and e^b, but if you want it between 10^a and 10^b, you can easily manipulate it by doing something like this: *loguniform*(*label*, $a * log_e(10)$, $b * log_e(10)$).

We can use these functions to create really complex search spaces. Let's look at an example of creating a search space:

```
from hyperopt import tpe, fmin, hp

space = hp.choice('classifier',[
      {'model': 'KNeighborsClassifier',
      'param': {'n_neighbors':
                hp.choice('n_neighbors',range(3,11)),
      'algorithm':hp.choice('algorithm', ['ball_tree',
      'kd_tree']),
```

```
    'leaf_size':hp.choice('leaf_size', range(1,50)),
    'metric':hp.choice('metric', ["euclidean", "manhattan",
                                   "chebyshev", "minkowski"
                  ])}
    },
    {'model': 'SVC',
     'param':{'C':hp.loguniform('C', -2*m.log(10),
     11*m.log(10)),
     'kernel':hp.choice('kernel',['rbf', 'poly', 'sigmoid']),
     'degree':hp.choice('degree', range(1,15)),
     'gamma':hp.loguniform('gamma', -9*m.log(10),
     3*m.log(10))}
    }
    ])
```

We start by choosing a classifier for our model using `hp.choice()`, one of KNN or SVM would be chosen. Once we choose the classifier, we can create distributions. In the preceding example, I chose to use *hp. choice*() for discrete distributions and *hp. lognormal*() for *C* and *gamma* in SVM. As mentioned earlier, we can manipulate function `hp.loguniform()` to use 10^x instead of e^x.

Note We need to provide labels to these hp functions. In the preceding code, I used the name of the hyperparameter itself to label the distribution.

Now that we have a search space, we'll work on the objective function:

```
from sklearn.datasets import load_digits
from sklearn.model_selection import train_test_split
from sklearn.neighbors import KNeighborsClassifier
from sklearn.svm import SVC
```

```
from hyperopt import tpe, fmin, hp

import math as m

digits = load_digits()
X_train, X_test, y_train, y_test = train_test_split(digits.data,
                                                    digits.target,
                                                    test_size=0.3)

logs = {'args':list(),
      'train_score': list(),
      'val_score': list()}

def objective_func(args):
      clf_func = args["model"]
      params = args["param"]

      clf = eval(clf_func)(**params)
      clf.fit(X_train, y_train)

      val_score = clf.score(X_test, y_test)
      train_score = clf.score(X_train, y_train)

      logs['args'].append(args)
      logs['train_score'].append(train_score)
      logs['val_score'].append(val_score)

      return -val_score

best = fmin(objective_func, space, algo=tpe.suggest,
max_evals=100)
```

Creating the objective function is easy. We extract the hyperparameters and pass them to the classifier. However, in this case, since our classifier is also a variable, we extract that as well. I have also created a log dictionary to save history. Run fmin() and the algorithm will start tuning. Refer to Figure 4-3-3, scatter plot of accuracy v/s trials.

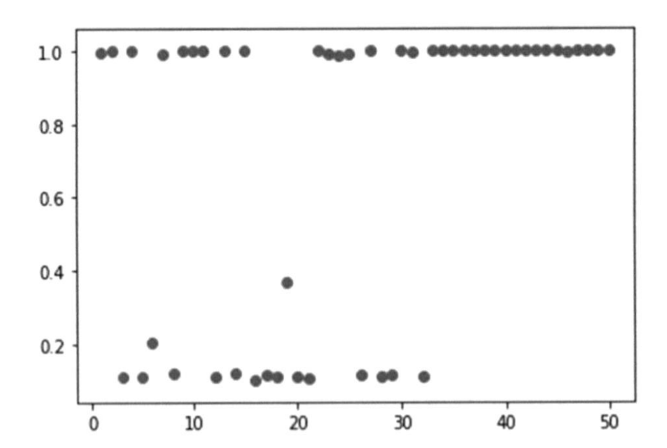

Figure 4-3-3. *Figure 4-3-3 shows the result of optimizing a support vector classifier for hyperparameters C, kernel, degree, and gamma on the digits dataset. We can see the score saturating at 0.99 after some 30 trials. Hence, the model is more certain about where to look to get the best hyperparameters*

Defining a neural network architecture using a Hyperopt search space can be a bit tricky where change in certain hyperparameters like the 'number of layers' change the total number of hyperparameters, because then we'll have to decide the number of nodes in each layer or whether or not layers use methods like batch normalization/dropout. So let's look at another interesting example that shows how we can define these kinds of awkward search spaces:

```
from hyperopt import hp, tpe, fmin
from keras.datasets import mnist
from keras.layers.core import Dense, Dropout, Activation
from keras.models import Sequential
from keras.utils import np_utils
import numpy as np

# load and preprocess the data
(x_train, y_train), (x_test, y_test) = mnist.load_data()
x_train = x_train.reshape(60000, 784)
```

```
x_test = x_test.reshape(10000, 784)
x_train = x_train.astype('float32')
x_test = x_test.astype('float32')
x_train /= 255
x_test /= 255
classes = 10
input_shape = 784
y_train = np_utils.to_categorical(y_train, classes)
y_test = np_utils.to_categorical(y_test, classes)

#logs
logs = {'model_summary':list(),
      'val_acc': list()}

def obj_func(args):

      #initializing the keras model
      model = Sequential()

      #defining first hidden layer
      model.add(Dense(units=args['units']['layer_units_1'],
                      input_shape=(input_shape, ),
                      name='layer_units_1'))

      #defining number of remaining hidden layer
      number_of_layers = len(args['units'])
      for layer in range(2, number_of_layers):
            model.add(Dense(units=args['units'][f'layer_units_
            {layer}'],
                            name=f'layer_units_{layer}'))
            model.add(Dropout(args['dropout'][f'dropout_p_
            {layer}'],
                         name=f'dropout_p_{layer}'))
```

```
            model.add(Activation(
                  activation=args['activation'][f'activation_
                  {layer}'],
                  name=f'activation_{layer}'))

      # adding last layer
      model.add(Dense(classes, name=f'layer_unit_{layer+1}'))
      model.add(Activation(activation='softmax',
                        name=f'activation_{layer+1}'))

      model.compile(loss='categorical_crossentropy',
                  metrics=['accuracy'], optimizer='adam')

     result = model.fit(x_train, y_train,
                        batch_size=2,
                        epochs=1,
                        verbose=3,
                        validation_split=0.2)

     validation_acc = np.amax(result.history['val_accuracy'])
     print(validation_acc)
     logs['model_summary'].append(model.summary())
     logs['val_acc'].append(validation_acc)

     return -validation_acc

def each_layer(number_of_layers):
     params = {'units': dict(),
              'dropout': dict(),
              'activation': dict()}
     number_of_nodes = [16,36,64,128,256,512]
     for layer in range(number_of_layers):
```

```
            params['units'][f'layer_units_{layer}'] = hp.choice(
                                                      f'layer_{number_of_
                                                      layers}_{layer}',
                                                      number_of_nodes)
            params['dropout'][f'dropout_p_{layer}'] = hp.uniform(
                                                     f'dropout_{number_of_
                                                     layers}_{layer}',
                                                     0, 0.8)
            params['activation'][f'activation_{layer}'] =
            hp.choice(
                                                 f'activation_{number_of_
                                                 layers}_{layer}',
                                                 ['relu', 'elu'])
        return params

# choice for number of layers
number_of_layers = [3, 5, 7, 9]
space = hp.choice('layers', [each_layer(n) for n in number_of_
layers])
best = fmin(obj_func, space, algo=tpe.suggest, max_evals=10)
```

In this particular case, first I decided the number of layers, and then I used a function, ***each_layer( )***, so that I can now iterate over the layers, deciding the number of nodes in each layer, the amount of dropout, and the choice of activation function. Note that I am using labels with `number_of_layers` and `layer` because I want the labels to be unique, similar to Dropout and Activation Functions. Now that the search space is created, we have to create the neural network in the objective function, which is pretty much straightforward. And now we can optimize hyperparameters for awkward searches.

Parallelizing Trials in TPE

We can exploit TPE even more by parallelizing the trials. Multiple candidates can be drawn at once from distribution $l(x)$, and these can be evaluated in parallel. By default, *fmin* is executed serially and uses the argument *trials* = *Trials*(), which uses a list. However, we can use *MongoTrials*() instead to evaluate these trials parallelly.

The first requirement, obviously, is to install MongoDB. After that, there are four simple steps to start the asynchronous optimization:

- When using *fmin*, pass trials as `MongoTrials()`.
- Start a visible MongoDB server.
- Execute the Python file.
- Run *hyperopt – mongo – worker*, which is a worker script placed in bin of your Python environment while installing Hyperopt.

Let's use MongoTrials() for the previous toy example of $f(a, b) = a^2 - b^2$:

```
from hyperopt import tpe, fmin, hp
from hyperopt.mongoexp import MongoTrials

def objective_func(args):
        a = args['a']
        b = args['b']
        f = a**2 - b**2
        return f

range_a = hp.uniform('a', -2, 3)
range_b = hp.uniform('b', -1, 2)

space = {'a': range_a,
             'b': range_b}
```

```
m_trials = MongoTrials("mongo://localhost:27017/foo_db/jobs",
exp_key="exp2")

best = fmin(objective_func, space, algo=tpe.suggest, trials=m_
trials, max_evals=1000)
```

In the preceding code we used MongoTrials(). The first step is to start a MongoDB server:

```
$ mongod --dbpath . --port 27017
```

By default, the port is 27017, but you can change it according to your need.

Now execute the previous Python script. In MongoTrials() we need to define the port and an exp_key, which you'll need to change in different runs if you are using the same database.

When you execute the script, it'll wait for mongo workers to start, which you can start with this command:

```
$ hyperopt-mongo-worker --mongo=localhost:1234/foo_db --poll-
interval=0.1
```

As previously mentioned, *hyperopt – mongo – worker* is a file stored in your $PATH (i.e., bin of the Python environment you are using). Here you need to give the *<host> < port > / < db – name>* and the poll interval checks work between every defined interval; if a job is found it'll start the computation.

As soon as you start the mongo workers, the sets of hyperparameters suggested by previous EI will be passed to the objective function and the process will start.

Note that worker is being executed in $PATH, and worker needs an objective function, so the Python script where the objective function is defined must be exported to $PATH.

And now you will get the asynchronous updates instead of serial ones. Alternatively, you can use Apache Spark for parallelization.

Hyperopt is designed to accommodate other surrogate functions like Gaussian process and random forest regression, but they are not implemented yet. But since Hyperopt is an open source library, I believe authors would certainly welcome these implementations. So go ahead and contribute to this amazing library for the greater good of the community.

Hyperopt-Sklearn

Hyperopt-sklearn[6] is a library[7] based on Hyperopt that uses Hyperopt for algorithm selection and hyperparameter tuning on scikit-learn algorithms.

The library can be a real time-saver because it creates its own search spaces for algorithms provided in scikit-learn. You can do end-to-end modeling, since it also provides algorithm selection and tuning options for data preprocessing (although not all scikit-learn algorithms are implemented yet).

The usage of hpsklearn is in sklearn style, implementing methods like `fit()`, `.score()`, and `.predict()` just like scikit-learn's Grid/Random Search. Providing a search space is optional though. Let's check out a few examples to understand it better.

```
from sklearn.datasets import load_boston
from sklearn.model_selection import train_test_split
from hpsklearn import HyperoptEstimator, any_regressor,
                                          any_preprocessing, svr
from hyperopt import tpe, hp
import math as m
```

[6] http://conference.scipy.org/proceedings/scipy2014/pdfs/komer.pdf
[7] https://github.com/hyperopt/hyperopt-sklearn

```
X, y = load_boston().data, load_boston().target
X_train, X_test, y_train, y_test = train_test_split(X,y, test_
size=0.2)

model = HyperoptEstimator(regressor=any_regressor('test1_reg'),
                    preprocessing=any_preprocessing('test1_
                    preprocessing'),
                    algo=tpe.suggest,
                    verbose=True,
                    max_evals=100)

model.fit(X_train, y_train, n_folds=3, cv_shuffle=True)

print(model.score(X_test, y_test))
print(mdoel.best_model())
```

Just like any other classifier/regressor in scikit-learn, we can use *HyperoptEstimator*, which means that even in your existing code, you need to change a single line to include this hyperparameter-tuning library. In the preceding code, we gave hpsklearn the freedom to choose any algorithm and set of hyperparameters and any preprocessing (normalization/standardization, etc.). However, we can restrict the tuning as well as preprocessing selection for certain algorithms. For example:

```
model = HyperoptEstimator(regressor=svr('test_svr'),
                          preprocessing=[],
                          algo=tpe.suggest,
                          verbose=True,
                          max_evals=100)
```

You can replace the initialization of the model in the previous code with this line to tune only on support vector regressor and use no preprocessing. Alternatively, you can change spaces for one or more hyperparameters like this:

```
space = {'C':hp.loguniform('C',-2*m.log(10),11*m.log(10)),
        'gamma':hp.loguniform('gamma',-9*m.log(10),3*m.log(10))
        }

model = HyperoptEstimator(regressor=svr('test_svr',
**space['param']),
                          preprocessing=[],
                          algo=tpe.suggest,
                          verbose=True,
                          max_evals=100)
```

Here, default search spaces for defined hyperparameters (C and gamma) will be overwritten by the custom search space. You can do the same with preprocessing.

Since the library supports algorithms like SVM, decision trees, KNN, and so on, you can use hpsklearn to get a baseline accuracy and use custom search spaces and try out different models to tune models further.

Hyperas

Yet another extremely useful open source hyperparameter optimization library, Hyperas[8] is a wrapper around Hyperopt for optimizing architecture of neural networks with Keras. This library is written in such a way that it saves you from creating complex search spaces for neural networks; instead, you can use simple Keras code with a little addition of ranges. Here's an example to help you understand the concept of Hyperas:

```
from hyperopt import Trials, STATUS_OK, tpe
from keras.datasets import mnist
from keras.layers.core import Dense, Dropout, Activation
```

[8] https://github.com/maxpumperla/hyperas

```
from keras.models import Sequential
from keras.utils import np_utils
import numpy as np

from hyperas import optim
from hyperas.distributions import choice, uniform

def data():
      # MNIST
      (x_train, y_train), (x_test, y_test) = mnist.load_data()
      x_train = x_train.reshape(60000, 784)
      x_test = x_test.reshape(10000, 784)
      x_train = x_train.astype('float32')
      x_test = x_test.astype('float32')
      x_train /= 255
      x_test /= 255
      classes = 10
      input_shape = 784
      y_train = np_utils.to_categorical(y_train, classes)
      y_test = np_utils.to_categorical(y_test, classes)
      return x_train, y_train, x_test, y_test, input_shape,
      classes

def create_model(x_train, y_train, x_test, y_test, input_shape,
classes):

      model = Sequential()
      model.add(Dense(units={{choice([8, 16])}},
                              input_shape=(input_shape,),
                              name='dense1'))

      layers = {{choice([2, 3, 4, 5, 6, 7, 8, 9, 10])}}
```

```
    for i in range(layers):
        model.add(Dense(units={{choice([32, 64, 256, 512,
        1024])}}))
        model.add(Dropout({{choice([0, 0.33])}}))
        model.add(Activation(activation={{choice(['relu',
        'elu'])}}))

    model.add(Dense(classes))
    model.add(Activation(activation='softmax'))

    model.compile(loss='categorical_crossentropy',
                  metrics=['accuracy'],
                  optimizer={{choice(['rmsprop', 'adam',
                  'sgd'])}})

    result = model.fit(x_train, y_train,
                  batch_size={{choice([4, 8, 16])}},
                  epochs=10,
                  verbose=3,
                  validation_split=0.2)

    validation_acc = np.amax(result.history['val_accuracy'])
    print('Test accuracy:', validation_acc)

    return {'loss': -validation_acc, 'status': STATUS_OK, 'model':
                                                         model}

best_run, best_model = optim.minimize(model=create_model,
                                      data=data,
                                      algo=tpe.suggest,
                                      max_evals=10,
                                      trials=Trials())
```

```
X_train, Y_train, X_test, Y_test, _, _ = data()
print("Test Score on Best Model:")
print(best_model.evaluate(X_test, Y_test))
print("Hyperparameter Set for best Model:")
print(best_run)
```

While using Hyperas, we need to create two functions. One function is for data loading. The other function is like an objective function that consists of a neural network and returns a loss score, the only difference being that the search space is not a parameter but instead data is passed as a parameter. The search space is defined as we write each hyperparameter in the network. And the *optim. minimize* () function from Hyperas starts the optimization.

In the first function, *data*(), we load the dataset (here MNIST); since the objective function will be iterating, we don't want to load data over and over. Everything you return from *data*() will be passed to *create_model*().

Next we define the objective function, where we define the neural network using Keras. After initializing `Sequential()`, we add layers one by one. In place of hyperparameters, we can give a range using this format: {{*'range'*}}. *'range'* is the distribution functions from Hyperas, which follow the same nomenclature as Hyperopt's. In Hyperas, we don't need to give labels, because it'll take the variables that are assigned to them as labels. However, if we are iterating over some hyperparameter range like in succeeding code,

```
for i in range(layers):
        model.add(Dense(units={{choice([32, 64, 256, 512, 1024])}}))
```

For instance, if there are three layers, the same hyperparameter will be chosen for all three because Hyperas makes a template of the Python code and sends the distributions to Hyperopt, which will consider it to be one hyperparameter, since it's written once. You have two alternatives to work around this problem: either you can go back and use Hyperopt as we did in last example in the "Search Space" section, or you can use *if...else* statement to add layers:

```
model.add(Dense({{choice([32, 64, 256, 512, 1024])}}))
model.add(Dropout({{uniform(0, 0.8)}}))
model.add(Activation({{choice(['relu', 'elu'])}}))

if {{choice(['one', 'two'])}} == 'two':
    model.add(Dense({{choice([32, 64, 256, 512, 1024])}}))
    model.add(Dropout({{uniform(0, 0.8)}}))
    model.add(Activation({{choice(['relu', 'elu'])}}))
```

If the number of chosen layers is two, only then will it create another layer.

Hyperas is a very simple and easy-to-use wrapper around Hyperopt, and you can use it to quickly tune your models, but if you want more flexibility, use Hyperopt, which works wonders even with the most complex search spaces.

In this chapter you learned how Bayesian hyperparameter optimization works and how you can use Hyperopt in your problems. These techniques can easily increase your time efficiency and optimize your resource utilization.

CHAPTER 5

Optuna and AutoML

We can now create an efficient model using the techniques that were discussed in the previous chapters. Bayesian optimization goes a long way in finding optimal hyperparameters. This chapter provides an overview of the Optuna framework and discusses further the role of hyperparameter optimization in automated machine learning. We'll use Optuna to create our own little AutoML script. And then we'll explore the Tree-based Pipeline Optimization Tool (TPOT), an AutoML tool that uses genetic programming to optimize machine learning pipelines.

Optuna

Like Hyperopt discussed in Chapter 4, Optuna[1] is open source library that uses Bayesian optimization. The underlying algorithms Optuna uses are the same as in Hyperopt, but the Optuna framework is much more flexible. Optuna can be easily used with PyTorch, Keras, scikit-learn, Apache MXNet, and other libraries. The API is very similar to Hyperopt's API, with a few changes. Let's dive into an example:

[1] "Optuna: A Next-Generation Hyperparameter Optimization Framework," T. Akiba, S. Sano, T. Yanase, T. Ohta, and M. Koyama, *KDD '19: Proceedings of the 25th ACM SIGKDD International Conference on Knowledge Discovery & Data Mining* (July 2019) 2623–2631.

T. Agrawal, *Hyperparameter Optimization in Machine Learning*,
https://doi.org/10.1007/978-1-4842-6579-6_5

```
from sklearn.datasets import load_digits
from sklearn.model_selection import train_test_split,
cross_val_score
from sklearn.neighbors import KNeighborsClassifier
from sklearn.svm import SVC

import optuna
from optuna.samplers import TPESampler

digits = load_digits()
X_train, X_test, y_train, y_test = train_test_split(digits.data,
                                                    digits.target,
                                                    test_size=0.3)

def objective_func(trial):

      classifier_name = trial.suggest_categorical("classifier",
                                    ["SVC", "RandomForest"])
      if classifier_name == "SVC":
            c = trial.suggest_loguniform("svc_c", 1e-2, 1e+11)
            gamma = trial.suggest_loguniform("svc_gamma", 1e-9,
            1e+3)
            kernel = trial.suggest_categorical("svc_kernel",
                                    ['rbf','poly','rbf',
                                    'sigmoid'])
            degree = trial.suggest_categorical("svc_degree",
            range(1,15))
            clf = SVC(C=c, gamma=gamma, kernel=kernel,
            degree=degree)

      else:
            algorithm = trial.suggest_categorical("algorithm",
                                    ['ball_tree', "kd_tree"])
            leaf_size = trial.suggest_categorical("leaf_size",
                                                range(1,50))
```

```
        metric = trial.suggest_categorical("metric",
                                     ["euclidean","manhattan",
                                      "chebyshev","minkowski"])
        clf = KNeighborsClassifier(algorithm=algorithm,
                                      leaf_size=leaf_size,
                                      metric=metric)

    clf.fit(X_train, y_train)
    val_acc = clf.score(X_test, y_test)

    return val_acc

study = optuna.create_study(direction='maximize',
sampler=TPESampler())
study.optimize(objective_func, n_trials=100)
best_trial = study.best_trial.value

print(f"Best trial  accuracy: {best_trial}")
print("parameters for best trail are :")
for key, value in study.best_trial.params.items():
    print(f"{key}: {value}")
```

In Figure 5-1-1, we can see the graph saturated around 1. Note that here the objective value is the validation accuracy.

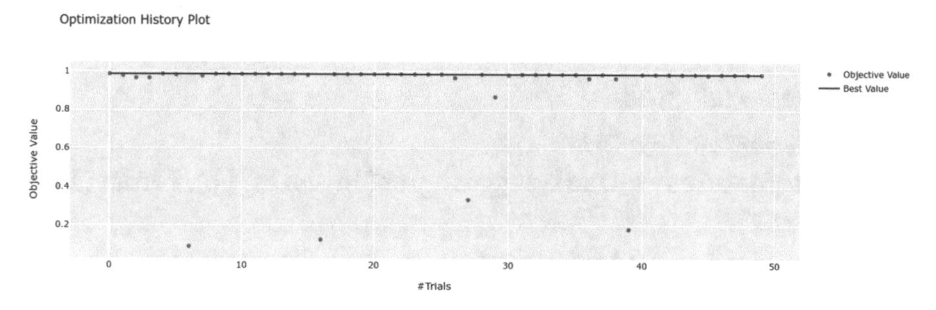

***Figure 5-1-1.** Plot showing the accuracy vs. trial on the first 50 trials of preceding code example*

If this example looks familiar, the reason is that we worked on the same problem in Chapter 4. The user interface of Optuna is quite similar to that of Hyperopt. We have to define an objective function that should return a score (loss/accuracy) which would be minimized/maximized.

In place of the `fmin()` function, we instantiate the `create_study()` function and optimize it. But one of the best features provided by Optuna is the capability to define the hyperparameter search range on the fly. Unlike Hyperopt, where we have to predefine the search space, in Optuna we define the search space in the objective function itself (something like what we did in Hyperas). Let's discuss some key aspects of Optuna.

Search Space

In Hyperopt and many other algorithms, we define search space using a dictionary. As mentioned, in Optuna, we define the search space on the fly. As you saw in Chapter 4, creating search spaces in neural networks is difficult with Hyperopt and Hyperas. In hyperparameters such as number of nodes that are dependent on the number of layers, Hyperas would use *if...else*. If a loop was used, Hyperas would choose the same number of nodes for all the layers. In Optuna we can provide the unique labels to each hyperparameter in a loop. For example:

```
n_layers = trial.suggest_int('n_layers', 1, 3)
layers = []

in_features = 28 * 28
for i in range(n_layers):
      out_features = trial.suggest_int('n_units_{}'.format(i),
      4, 128)
      layers.append(Linear(in_features, out_features))
      layers.append(ReLU())

      in_features = out_features
```

Optuna makes it so much easier to deal with this problem. And Optuna provides the same distributions as Hyperopt. The following are some of the commonly used distributions:

- *Categorical distribution*: `trial.suggest_categorical()` is used for selecting discrete values from a list, which is the same as `hp.choice()` in Hyperopt.
- *Uniform distribution*: `trial.suggest_uniform()` is used for a random distribution on a uniform scale, which is similar to `hp.uniform()`.
- *LogUniform Distribution*: trial.suggest_loguniform(label, low, high) is used for a loguniform scale. However, in Hyperopt `hp.loguniform(label, low, high)` returns a distribution between *exp*(*low*) and *exp*(*high*). In Optuna, a loguniform distribution between *low* and *high* is returned.

And there are more distributions we can use as per our need.

Underlying Algorithms

In addition to tree-structured Parzen estimator and random search, Optuna provides successive halving and HyperBand, which is an advantage over Hyperopt. We have already discussed HyperBand in Chapter 3. Here's how we can use it:

```
import optuna

# define the objective function

study = optuna.create_study(pruner=optuna.pruners.
HyperbandPruner())
study.optimize(objective, n_trials=20)
```

Visualization

Optuna provides elegant visualization. Figure 5-1-1 is generated by Optuna using *optuna. visualization. plot _ optimization _ history(study)*. You can pass the `study` object and it'll create a graph. You can point on each scatter point and observe different hyperparameters used. There are even more visualization options.

Callbacks, an argument in *study.optimize()* which invokes callback after each trial. Using this you can visualize progress on dashboards like *tensorboardX* in real time.

Callbacks work something like this:

```
import optuna
from optuna.samplers import TPESampler

def log(study, trial):
      print(f"Trial No.={trial.number}, HP_Set={trial.params}, \
      Score={trial.value}")
      print(f"Best Value ={study.best_value}")

# def objective_func()

study = optuna.create_study(sampler=TPESampler())
study.optimize(objective_func, n_trials=100, callbacks=[log])
It's really easy to work with. You can write these results in
the dashboard and they'll get updated after each trial.
```

Distributed Optimization

Just like Hyperopt, Optuna supports distributed optimization, but working with Optuna's implementation is easier than working with Hyperopt's implementation. Let's how we can configure it:

```
study = optuna.create_study(study_name='distributed_test',
                            storage='database_url',
                            load_if_exists=True)
```

Define database url while instantiating '*create _ study*()'. And set 'load_if_exits=True' this instead of creating a new study would look for a previous study named "distributed_test". That way, every time a worker starts, it won't create a new study but instead look for the existing one, and thus won't start training from scratch.

For a comprehensive comparison of Hyperopt and Optuna, refer to the following article by Jakub Czakon, Senior Data Scientist at Neptune.ai: "Optuna vs Hyperopt: Which Hyperparameter Optimization Library Should You Choose?"[2].

Now, let's explore an example of how we can optimize the hyperparameters of a neural network using Optuna. We'll be working with the MNIST dataset and Keras. We start by importing libraries and split the data to train and test set.

```
from keras.datasets import mnist
from keras.layers.core import Dense, Dropout, Activation
from keras.models import Sequential
from keras.utils import np_utils
import numpy as np

import optuna
from optuna.samplers import TPESampler

(x_train, y_train), (x_test, y_test) = mnist.load_data()
x_train = x_train.reshape(60000, 784)
x_test = x_test.reshape(10000, 784)
x_train = x_train.astype('float32')
```

[2]https://neptune.ai/blog/optuna-vs-hyperopt

```
x_test = x_test.astype('float32')
x_train /= 255
x_test /= 255
classes = 10
input_shape = 784
y_train = np_utils.to_categorical(y_train, classes)
y_test = np_utils.to_categorical(y_test, classes)
x_train, y_train, x_test, y_test, input_shape, classes

def log(study, trial):

      print(f"Trial No.={trial.number}, HP_Set={trial.params}, \
            Score={trial.value}")
      print(f"Best Value ={study.best_value}")

def objective_func(trial):

      model = Sequential()

      hidden_layer_unit_choice = [32, 64, 256, 512, 1024]

      hidden_layers = trial.suggest_int('hidden_layers', 1, 6)

      model.add(Dense(units=trial.suggest_categorical('layer1',
      [8, 16]),
                  input_shape=(input_shape, ),
                  name='dense1'))

      model.add(Activation(activation=trial.suggest_categorical(
                                                 f'activation1',
                                                        ['relu',
                                                        'elu'])))
```

```
for i in range(1, hidden_layers):

    model.add(Dense(units=trial.suggest_categorical(
                                f'layer{i+1}',
                                hidden_layer_
                                unit_choice)))
    model.add(Dropout(trial.suggest_uniform(
                                f'dropout{i+1}',
                                0, 0.8)))
    model.add(Activation(
                   activation=trial.suggest_
                   categorical(
                              f'activation{i+1}',
                                        ['relu',
                                        'elu'])))

model.add(Dense(classes))
model.add(Activation(activation='softmax'))

model.compile(loss='categorical_crossentropy',
          metrics=['accuracy'],
          optimizer=trial.suggest_categorical('optimizer',
          ['rmsprop', 'adam', 'sgd']))

result = model.fit(x_train, y_train,
              batch_size=4,
              epochs=1,
              verbose=3,
              validation_split=0.2)

validation_acc = np.amax(result.history['val_accuracy'])
print('Validation accuracy:', validation_acc)

return validation_acc
```

Define the objective function. In the preceding code, we see hyperparameters beign selected for each layer on the fly, just by giving unique names to labels.

And lastly, we start the optimization:

```
study = optuna.create_study(direction='maximize',
sampler=TPESampler())
study.optimize(objective_func, n_trials=50, callbacks=[log])
best_trial = study.best_trial.value

print(f"Best trial  accuracy: {best_trial}")
print("parameters for best trail are :")
for key, value in study.best_trial.params.items():
   print(f"{key}: {value}")
```

This example optimizes for 50 trials only, and Figure 5-1-2 shows the accuracy graph.

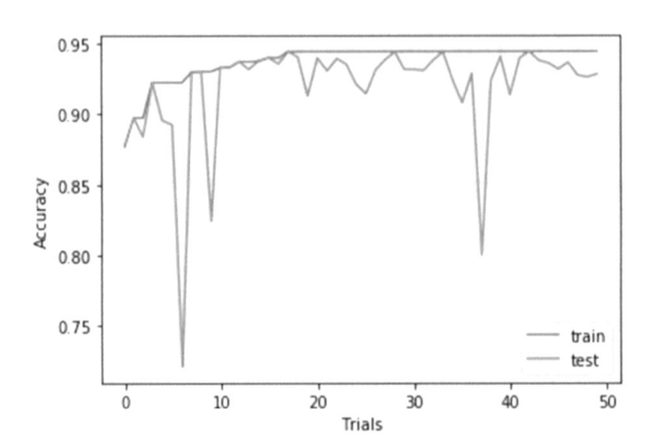

Figure 5-1-2. *Accuracy vs. trials for the previous code example*

In such a huge search space 50 trials are less (increase the number of trials for better results), but we can see the training score increasing. Note that the test score is independent, and the objective function is using validation accuracy to optimize hyperparameters.

Optuna is a young library, with a lot of work still in progress, but it's promising.

Automated Machine Learning

The high complexity of machine learning demands that only machine learning experts build models. Machine learning models are task-specific, where each model requires a lot of work. To provide machine learning to the masses, machine learning experts need a method to create off-the-shelf models. This is where automated machine learning (AutoML) steps in.

Machine learning is automated when it creates the complete pipeline and gives us a deployable model on its own. To create a complete pipeline, we need to use several algorithms, from preprocessing to creating a machine learning model. All these algorithms have their own hyperparameters that need to be optimized. Here hyperparameter optimization plays a huge role. The steps and algorithms for preprocessing are described in Appendix I.

Now we'll quickly build our own AutoML modules in subsequent sections using Optuna and TPOT, which would be able to handle almost any kind of dataset for classification.

Building Your Own AutoML Module

This example provides instructions for building a basic module that you can subsequently experiment with and add more algorithms. For use in real-world problems, there are many optimized AutoML libraries, which we'll discuss later. We'll work on the Titanic dataset, which is messy, but our code will handle all the cleaning and preprocessing.

Let's start by importing everything and loading the dataset[3]:

```
import pandas as pd
import numpy as np
import impyute as impy

import optuna
from optuna.samplers import TPESampler

from sklearn.model_selection import train_test_split
from sklearn.preprocessing import LabelEncoder, MinMaxScaler, \
        StandardScaler
from sklearn.impute import SimpleImputer
from sklearn.feature_selection import SelectKBest, \
          chi2, f_classif, mutual_info_classif
from sklearn.svm import SVC
from sklearn.neighbors import KNeighborsClassifier

data = pd.read_csv("./titanic/train.csv")

y = data['Survived']
X = data.drop('Survived', axis=1)
```

Now we'll address the part where we are not optimizing the hyperparameters. We'll define the outside objective function because it does not make sense to iterate the same process that is not to be optimized.

```
def label_encode_column(col):
      nans = col.isnull()
      nan_lst = []
      nan_idx_lst = []
      label_lst = []
      label_idx_lst = []
```

[3] https://www.kaggle.com/c/titanic

```
    for idx, nan in enumerate(nans):
        if nan:
            nan_lst.append(col[idx])
            nan_idx_lst.append(idx)
        else:
            label_lst.append(col[idx])
            label_idx_lst.append(idx)

    nan_df = pd.DataFrame(nan_lst, index=nan_idx_lst)
    label_df = pd.DataFrame(label_lst, index=label_idx_lst)

    label_encoder = LabelEncoder()
    label_df = label_encoder.fit_transform(label_
    df.astype(str))
    label_df = pd.DataFrame(label_df, index=label_idx_lst)
    final_col = pd.concat([label_df, nan_df])

    return final_col.sort_index()

for column_name in X.columns:
    if str(X[column_name].dtype) == 'object':
        X[column_name] = label_encode_column(X[column_name])
        if len(X[column_name].unique()) > len(X)/3:
            X = X.drop(column_name, axis=1)
```

We are using the function `label_encode_column()`, but what's wrong with just LabelEncoding()? LabelEncoding() also labels the NaN values, but we want to impute those later. So function takes each column, separates the NaN and other values, and labels them while saving their index position. It returns a sorted dataframe, an example of which is shown in Figure 5-2-1.

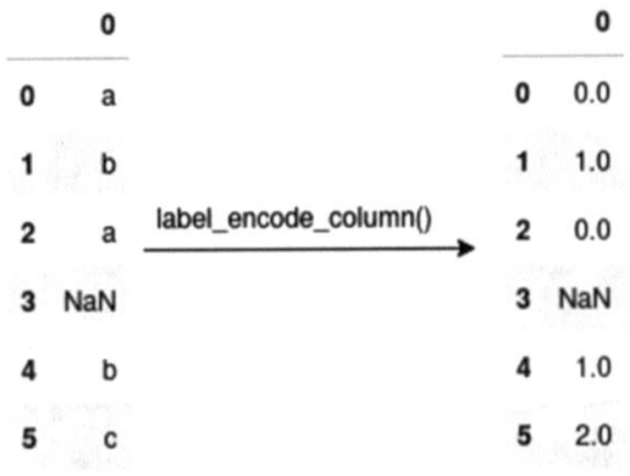

	0
0	a
1	b
2	a
3	NaN
4	b
5	c

label_encode_column() →

	0
0	0.0
1	1.0
2	0.0
3	NaN
4	1.0
5	2.0

Figure 5-2-1. *Label Encoder encodes all values except NaNs*

So after we label encode all the columns, we drop those with a high number of unique values. You can alternatively include label encoding in the objective function if you want one-hot encoding as an option. Next, we are going to define some functions to be used in the objective function:

```
def mice_imputer(data):
        data = data.to_numpy()
        imputed_data = impy.mice(data)
        imputed_data = pd.DataFrame(imputed_data)
        return imputed_data

def mean_imputer(data):
        imputer = SimpleImputer(strategy='mean')
        imputed_data = imputer.fit_transform(data)
        imputed_data = pd.DataFrame(imputed_data)
        return imputed_data
```

The preceding code parameterizes two imputers, Mean Imputation and MICE Imputation. Refer to Appendix I for instructions on including more options.

```
def feature_selector(X, y, k, algo="f_classif"):
        kbest = SelectKBest(eval(algo), k)
        X = kbest.fit_transform(X, y)
        X = pd.DataFrame(X)
        return X
```

We are going to select 'k' number of best features, which is also going to be a hyperparameter. Another hyperparameter is selecting the feature selection algorithm. We have 'f_classif', 'chi2', and 'mutual_info_classif'. The last of the preprocessing steps is scaling. We select between normalization and standardization.

```
def scaling(data, scaler="min_max"):
      if scaler=='min_max':
            scaled_data = MinMaxScaler().fit_transform(data)
      else:
            scaled_data = StandardScaler().fit_transform(data)
      scaled_data = pd.DataFrame(scaled_data)
      return scaled_data
```

And now we define the objective function:

```
def objective_func(trial):

      try:
            # imputation
            imputer = trial.suggest_categorical('impyter',
            ['mice', 'mean'])
            if imputer=='mice':
                  imputed_X = mice_imputer(X)
            else:
                  imputed_X = mean_imputer(X)

            # feature selection
            fea_slct = trial.suggest_categorical('fea_slct',
                                                 ['chi2',
                                                 'f_classif',
                                                 'mutual_info_
                                                 classif'])
            no_feature_cols = trial.suggest_int('k',
            3, len(X.columns))
```

```
        selected_features = feature_selector(imputed_X, y,
                                             no_feature_cols,
                                             fea_slct)

        # scaling
        scaler = trial.suggest_categorical('scaler',
                                  ['min_max', 'standard'])
        scaled_X = scaling(selected_features)

    except:
        return 0.0

    # instantiating machine learning algorithm
    classifier_name = trial.suggest_categorical("classifier",
                                              ["SVC",
                                              "RandomForest"])

    if classifier_name == "SVC":
        c = trial.suggest_loguniform("svc_c", 1e-2, 1e+11)
        gamma = trial.suggest_loguniform("svc_gamma",
        1e-9, 1e+3)
        kernel = trial.suggest_categorical("svc_kernel",
                                  ['rbf','poly','rbf',
                                  'sigmoid'])
        degree = trial.suggest_categorical("svc_degree",
        range(1,15))
        clf = SVC(C=c, gamma=gamma, kernel=kernel,
        degree=degree)
    else:
        algorithm = trial.suggest_categorical("algorithm",
                                            ['ball_tree',
                                            "kd_tree"])
        leaf_size = trial.suggest_categorical("leaf_size",
        range(1,50))
```

```
        metric = trial.suggest_categorical("metic",
                                           ["euclidean",
                                            "manhattan",
                                            "chebyshev",
                                            "minkowski"])
        clf = KNeighborsClassifier(algorithm=algorithm,
                                   leaf_size=leaf_size,
                                   metric=metric)

    # fit the model
    clf.fit(scaled_X, y)
    val_acc = clf.score(scaled_X, y)

    return val_acc

study = optuna.create_study(direction='maximize',
sampler=TPESampler())
study.optimize(objective_func, n_trials=100)
best_trial = study.best_trial.value

print(f"Best trial  accuracy: {best_trial}")
print("parameters for best trial are :")
for key, value in study.best_trial.params.items():
    print(f"{key}: {value}")
```

The first part of the objective function consists of all the preprocessing steps, where we are using `try...except`. We are using the functions defined before. In some cases, when there is a mismatch in algorithm and hyperparameter, or if the data processed by the previous step is not acceptable by the next step, you might encounter an error. For instance, some of the feature selection algorithms don't work on negative values, but even if you are careful with the dataset, the imputer might impute a NaN with some negative value. In that case, just return zero, the least possible accuracy value.

We then define the choice for classifiers as we did before (you can add more classifiers). The idea of AutoML is to create a generic code that gives you the best possible pipeline for the dataset. You can train any tabular dataset for classification to get the best possible pipeline for the previous set of steps and choices of algorithm without writing another line of code.

As the complexity of the search space is increased, we need to perform more trials over the dataset. The more trials, the better the results we'll achieve. To reduce the time, you can distribute the optimization process, as described next.

TPOT

The Tree-based Pipeline Optimization Tool (TPOT) is an AutoML framework that uses genetic programming to optimize the machine learning pipeline.

As we have discussed, data preprocessing typically consists of data cleaning (label encoding, dropping unimportant columns, and scaling), which is something we must take care of beforehand. The more complex tasks, such as feature selection, feature reduction, and feature construction, are handled by TPOT. It further selects the best model with the best set of hyperparameters. Figure 5-2-2 shows the features that are automated by TPOT.

Genetic algorithms are slow but excel at finding the best route for a given dataset. We would need to train for a long time before reaching the best set of hyperparameters.

Understanding TPOT first requires understanding what genetic algorithms are. As the name suggests, the concept of genetic algorithms is derived from Darwin's theory of natural selection. A genetic algorithm continuously evolves, by selecting the best algorithm. For the best algorithm it goes to its children doing some random modifications in hyperparameters and evaluating models to find the best fit.

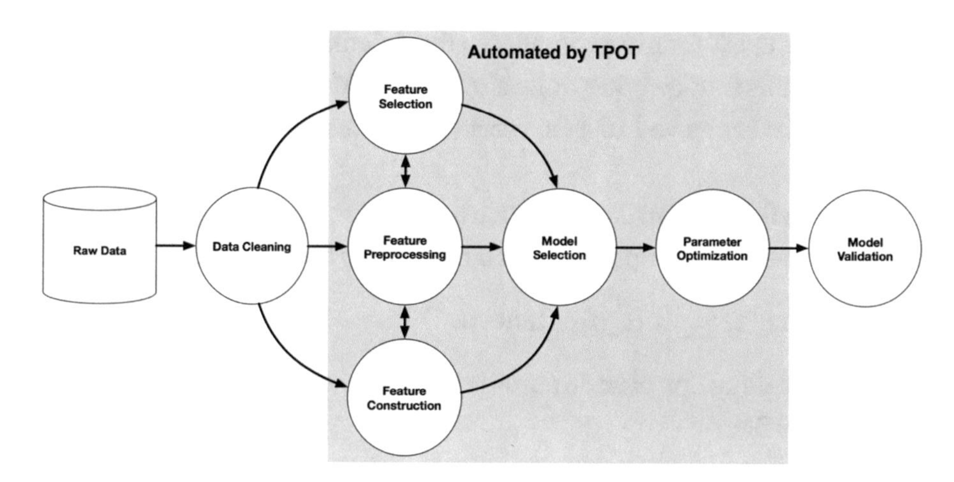

Figure 5-2-2. *TPOT covers around 80% of the job*

Now we'll look at an example of how TPOP works, using TPOT for the Iris dataset, since it's already a clean dataset:

```
from tpot import TPOTClassifier
from sklearn.datasets import load_iris
from sklearn.model_selection import train_test_split
import numpy as np

iris = load_iris()
X_train, X_test, y_train, y_test = train_test_split(
                                        iris.data.astype
                                        (np.float64),
                                        iris.target.astype
                                        (np.float64),
                                           test_size=0.25
                                           )
```

Loading the dataset and splitting it to train and test.

```
tpot = TPOTClassifier(generations=5, population_size=50,
verbosity=2)
tpot.fit(X_train, y_train)
```

We train TPOT for five generations; the default is 100 generations. The interface of this framework is quite similar to scikit-learn.

And voila! I achieved 100% accuracy on the test set on the trained model.

To make things even easier, we have an option to export the selected pipeline to a Python script, which can then easily be deployed.

```
tpot.export('tpot_iris_pipeline.py')
```

The preceding line of code auto-generates the following tpot_iris_pipeline.py file:

```
import numpy as np
import pandas as pd
from sklearn.model_selection import train_test_split
from sklearn.neural_network import MLPClassifier

# NOTE: Make sure that the outcome column is labeled 'target'
in the data # file

tpot_data = pd.read_csv('PATH/TO/DATA/FILE', sep="COLUMN_
SEPARATOR",
                        dtype=np.float64)
features = tpot_data.drop('target', axis=1)
training_features, testing_features, training_target, testing_
target = \
           train_test_split(features, tpot_data['target'],
                            random_state=None)

# Average CV score on the training set was: 0.9913043478260869
exported_pipeline = MLPClassifier(alpha=0.1, learning_rate_
init=0.1)

exported_pipeline.fit(training_features, training_target)
results = exported_pipeline.predict(testing_features)
```

For more time we optimize, the better the results we'll get.

There are many more libraries for AutoML, such as auto-sklearn, H2O AutoML, and AutoKeras. AutoML not only is beneficial to data scientists to accelerate the process but also has enabled people with no knowledge of coding to use machine learning. AutoML certainly holds a great position in the future of artificial intelligence.

As you've witnessed in this chapter, hyperparameter optimization plays a huge role in AutoML, enabling us to generate the best machine learning models in the least amount of time.

To conclude this book, I wish you success as you optimize your hyperparameters using more advanced methods like Bayesian optimization and pruning algorithms instead of manually tuning or grid search. Happy optimization!

Appendix I

The purpose of this appendix is to help you brush up on basic machine learning concepts and look at ways to evaluate models.

Data Cleaning and Preprocessing

Clearly, data cleaning and preprocessing is the most important task when making a machine learning model. When dealing with real-life data, you will find nonnumerical columns, missing values, outlier data points, unwanted features, and so forth.

Before you start preprocessing data, you must carefully look and understand the dataset, and understand the meaning of each column if possible.

The following sections address all the data cleaning and preprocessing problems that you may encounter and discuss algorithms that can be used to solve them. After that, we'll explore how to apply them to a real-world dataset.

Dealing with Nonnumerical Columns

Algorithms understand numbers but not strings. If a column consists of strings, we must change them to a numbers. But what if each point in the column is a unique string (for example, a dataset containing unique names)? In that case, the column must be dropped, so it's important to look at the dataset carefully.

In this section we'll look at some methods to convert strings to numbers.

T. Agrawal, *Hyperparameter Optimization in Machine Learning*,
https://doi.org/10.1007/978-1-4842-6579-6

Label Encoding

We have few unique strings in a column, when we convert them to labels, for example, we have these unique string, "a", "b", "c" in repetition we convert them to 0, 1, 2 respectively. Now every occurrence of string "a" will be replaced by number 0. In Table A1-1, Col 1 represents a feature column and Col 1' represents its replacement label encoding that is understandable to algorithms. This process of converting string to numbers (labels) is called *Label Encoding*.

Table A1-1. *Col 1 Replaceable by Col 1' Through Label Encoding*

Col 1	Col 1'
a	0
b	1
b	1
c	2
a	0

Label encoding introduces one problem, though: in certain contexts, such as categorical data, representing strings with numbers suggests a prioritization, ranking, or order of the strings where no such correspondence exists. That is, assigning 0, 1, and 2 to a, b, and c doesn't mean, for example, that a precedes b in order and b precedes c. To overcome this problem, we can use one hot encoding.

One-Hot Encoding

One-hot encoding is another method to convert categorical data to numeric. Here we split one column into multiple columns, where the number of new columns equals the number of unique strings in the categorical data. If we have three categories, “a”, “b”, and “c”, each occurrence of “a” will be assigned as [1, 0, 0], “b” as [0, 1, 0], and c as [0, 0, 1]. 1 being value exist in newly created column and 0 means it doesn’t. From Col 1 in Table A1-1, if we one-hot encode the categorical data, the result is as shown in Table A1-2.

Table A1-2. *One-Hot Encoding of Col 1 of Table A1-1*

Col 1' a	Col 1' b	Col 1' c
1	0	0
0	1	0
0	1	0
0	0	1
1	0	0

Theoretically, one-hot encoding is superior to label encoding, but the number of columns will increase as the number of categories increases. When we actually compute the machine learning algorithms, the complexity increases exponentially with increase in features. So there’s our trade-off.

Missing Values

Now that we have handled the categorical string columns, let’s handle the missing values.

Real-life datasets can contain missing values, for various reasons. Often these missing values are identified as NaN, as shown in Table A1-3, or blank cells or even empty strings (""). There are many ways you can deal with missing values, as described next.

Table A1-3. Variables with Missing Values (NaN)

C1	C2	C3	C4
21.1	1	4562	198.0
NaN	NaN	2433	183.6
24.3	0	NaN	211.7

Drop the Rows

Drop the rows containing missing values. Before doing that, you must check though, if you have enough data points left (to train a machine learning model) even after dropping those rows with missing values. Or if points in a particular feature are mostly missing, you can drop that too.

Mean/Median or Most Frequent/Constant

You can fill the missing values with the mean or median over values in same column. Although this is the easiest method to deal with missing values, it is quite inaccurate. Also, it can be used only on continuous features and not on categorical ones. For instance, if you use mean imputation for C2 in Table A1-3, 0.5 is the mean value, but C2 might consist of only 1s and 0s.

Filling with most frequent or constant values from the same column would work with categorical features, but it can introduce a bias to the data.

Neither of these methods addresses the correlation between features.

Imputation Using Regression or Classification

You can use an algorithm like support vector machine (SVM) or K-nearest neighbor (KNN) to predict the missing values, using the features that don't have missing values. For example, in Table A1-3, likes of C4 can be used as features and C1 or C2 can be termed target columns. Predictive algorithms can learn the relation between C1 and C4 or C2 and C4 and predict NaNs using regression and classification, respectively.

Multivariate Imputation by Chained Equations[1]

The previous methods wouldn't work if you have missing values in all the columns.

In this process of MICE imputation, missing values are filled multiple times to complete the dataset. Let's go step by step through how it works:

1. It calculates and imputes using the mean imputation for each missing position, which can be termed as "placeholders."
2. Placeholders for one of the features [F] is set to missing; that is, all the values we imputed using mean in feature F are set to missing.
3. F is set as a target column and rest are set as feature columns, and F is regressed on all the other features.
4. We impute the missing values in F, and F is now used as a feature for other features to be imputed.
5. This cycle repeats until all the features are imputed.
6. The whole cycle, steps 1 to 5, is repeated for imputing all the features again and again. The number of cycles is a factor that we decide and is based on experimentation.

[1]"Multiple Imputation by Chained Equations: What Is It and How Does It Work?" M. Azur, E. Stuart, C. Frangakis, and P Leaf. *Int J Methods Psychiatr Res.* 20(1) (2011) 40–49. doi:10.1002/mpr.329

Outlier Detection

In any given dataset, sometimes a few observations deviate from most other observations, creating a biased weightage in their favor; known as *outliers*, they must be removed in order to eliminate unwanted bias. For example, the top image in Figure A1-1 shows a normal (Gaussian) distribution, which has an approximate mean of 5 on both the x axis and y axis. The bottom image in Figure A1-1 shows a threshold ellipse drawn to define points under the distribution and outliers. Point (-0.8,4.2) is an obvious outlier, and some of the other points that lie outside the threshold circle can be assumed to be outliers since they don't follow the general pattern of distribution.

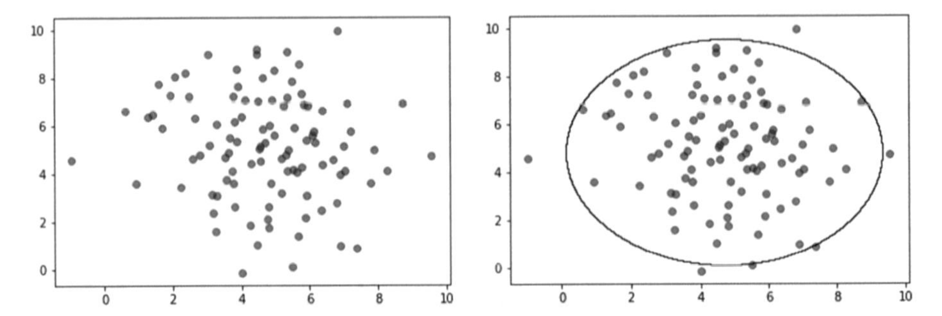

***Figure A1-1.** A Gaussian distribution with mean 5 and standard deviation 2.5 (top) and identification of outliers by drawing an ellipse (bottom)*

This distribution could be visualized because it is two dimensional; however, we would usually witness datasets with a large number of features, which means a greater number of dimensions that we won't be able to visualize. So we rely on algorithms to detect these outliers. I'll discuss a few of them further.

Z-score

Let's assume the data is Gaussian, hence making a bell curve. Intuitionally, z-score tells us how far the data point is from the mean position (where most of the data points lie). The formula for calculating z-score is

$$z = (x - \mu) \div \sigma$$

where x is the data point, μ is the mean of the sample data, and σ is the standard deviation of the sample. Now we set a certain threshold of z, and accordingly eliminate the data points that are outliers.

But what if the data is not Gaussian? We can normalize the data or we can use log transformation or Bob Cox transformation in case of skewed columns.

A few limitations with outlier detection based on z-score are that it can't be used on nonparametric data and the number of dimensions should be less.

Density-Based Spatial Clustering of Applications with Noise

Density-based spatial clustering of applications with noise (DBSCAN) is a clustering algorithm that clusters on the basis of density of points. Points lying in low-density areas can be marked as outliers. See Figure A1-2.

The forming of clusters depends on two factors: ε and *min_samples*. ε is the minimum distance between two points that can be considered as neighbor. If ε is too low, the result will be few neighbors and too many clusters, and no dense region will be formed; however, if ε is too high, the result will be one of the clusters consisting of most of the points. *min_samples* is the minimum number of points required to make a cluster.

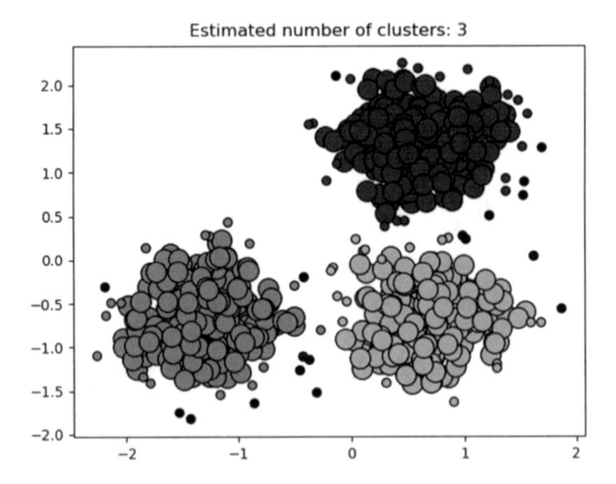

Figure A1-2. *An example of DBSCAN forming clusters to find which data points lie in low-density area*[2]

Feature Selection

Features are the key to map a relationship between data points and the target value. If some of those features are corrupt and are independent of the target values, they are no help in mapping that relationship. Therefore, we hunt them out and remove them from the dataset. There are two different types of algorithms, univariate and multivariate, that help us to hunt for these features.

- *Univariate algorithms* try to find the relationship between each feature (independently of other features) and the target columns. If the relationship is strong, we keep the feature; otherwise, we discard it.
- *Multivariate algorithms* find the dependency across the features. We get the score for all the features and select the best ones.

[2] https://scikit-learn.org/stable/auto_examples/cluster/plot_dbscan.html

Next we'll look at some statistical methods and algorithms that help us do so.

F-Test

F-test creates a comparison between two models, each of which is first created by a constant and another by a constant and a feature. Then the F-test finds out whether the relationship between feature and target actually means something. An F-test is only able to capture the linear dependency between feature and target. This problem is solved by the next method we'll discuss.

Mutual Information Test

As the name suggests, this score finds out the mutual dependency between two variables (here a feature and a target column). Mutual information calculates the amount of information we get about one variable (feature) given another (target). Mutual information between two variables *X* and *Y* can be stated as follows:

$$\text{Mutual Info}(X, Y) = H(X) - H(X \mid Y)$$

where H(X) is the entropy of X, and H(X | Y) is the conditional entropy of X given Y.

This captures even the nonlinear relation between X and Y. As you can see in Figure A1-3, the F-score captures the linear relationship but ignores the randomness in the feature, while the mutual information (MI) score considers the nonlinear relationship very well.

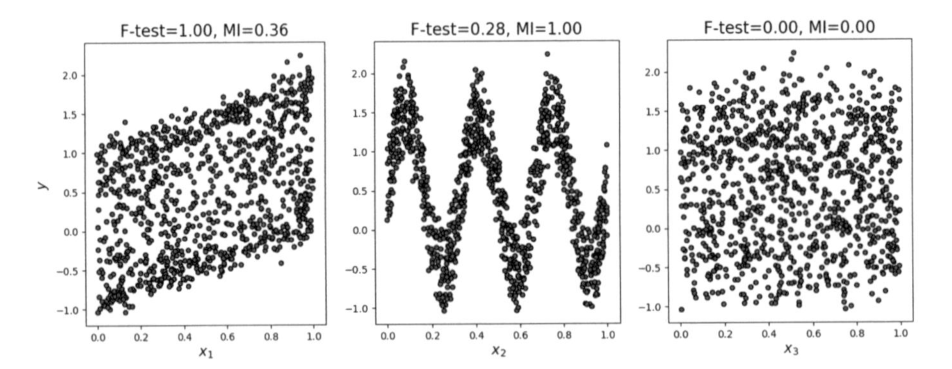

Figure A1-3. *From left to right, three data distributions and their respective F-test score and mutual information score written on the top*[3]

Recursive Feature Selection

Whereas the previous two methods are univariate, recursive feature selection can be classified as multivariate. Here we train a complete model and the model gives the weightage to each variable, called *coefficients*. Features are then sorted according to these values of coefficients, and the features with low coefficients are removed.

This pretty much sums up the cleaning and preprocessing, there's so much more that can be done according to the problem and dataset.

Applying the Techniques

This section demonstrates step by step how to apply some of the data cleaning and preprocessing methods discussed thus far to a very generic dataset, the Titanic dataset.

[3] https://scikit-learn.org/stable/auto_examples/feature_selection/plot_f_test_vs_mi.html

First, import the following libraries:

```
import pandas as pd
import numpy as np
from sklearn.preprocessing import LabelEncoder
import impyute as impy
from scipy import stats
```

We'll use these libraries as follows:

- Pandas to load the Titanic dataset
- Numpy to deal with array (operating on vectors) and perform some calculations
- Scikit-learn to help in the label encoding
- Impyute is a library which implements different implementation algorithm and is really easy to use
- Scipy to help with some more calculations

```
# Read the data
data = pd.read_csv("./titanic/train.csv")
print(data.head())
```

Our data looks something like Table A1-4.

Table A1-4. Output of data.head(), Displaying Five Rows by Default

PassengerId	Survived	Pclass	Name	Sex	Age	SibSp	Parch	Ticket	Fare	Cabin	Embarked
1	0	3	Braund, Mr. Owen Harris	male	22.0	1	0	A/5 21171	7.2500	NaN	S
2	1	1	Cumings, Mrs. John Bradley (Florence Briggs Th...	female	38.0	1	0	PC 17599	71.2833	C85	C
3	1	3	Heikkinen, Miss. Laina	female	26.0	0	0	STON/O2. 3101282	7.9250	NaN	S
4	1	1	Futrelle, Mrs. Jacques Heath (Lily May Peel)	female	35.0	1	0	113803	53.1000	C123	S
5	0	3	Allen, Mr. William Henry	male	35.0	0	0	373450	8.0500	NaN	S

RMS *Titanic* was a ship that sank after hitting an iceberg in 1912, taking with her over 1500 passengers. However, a few hundred survived as well. Many factors contributed to whether or not any particular passenger survived or died. For instance, women and children were given preference for boarding lifeboats. And if a woman was traveling first class, she would have been among the first to board a lifeboat, increasing her chance of survival.

Now we have a dataset to analyze, given different variables like Age, Sex, Passenger Class, Fare, etc. we have to train a model to predict whether a passenger survived or not. We have data for 891 passengers.

One of the most important things you can do is look and understand the data. We don't need a feature selection algorithm to decide if features like Name or Passenger ID are important or not. All the names are unique and thus won't map any relationship with the target; the same is true with passenger IDs. So, based on mere observation of the dataset, you can decide which features to keep and which to discard. For example:

```
data.drop(["Name", "Cabin", "Ticket", "PassengerId"], axis=1,
inplace=True)
```

`data.drop()` discards the list of features, and `axis=1` implies to drop columns of these names.

Let's now check which columns contain missing values. The following command shows us the column-wise number of missing points:

```
>>> data.isna().sum()

Survived      0
Pclass        0
Sex           0
Age         177
SibSp         0
Parch         0
Fare          0
Embarked      2
dtype: int64
```

We see that Embarked and Age are two columns with missing values. Whereas Age has too many data points missing, Embarked has only two, and it would be fine if we drop those two data points completely:

```
data = data.dropna(subset=["Embarked"])
```

Before starting the imputation, we'll label encode the categorical columns represented by the strings "Sex" and "Embarked":

```
L = LabelEncoder()
data["Sex"] = L.fit_transform(data["Sex"])
data["Embarked"] = L.fit_transform(data["Embarked"])
```

Now, our dataset will look something like Table A1-5.

Table A1-5. *Table A1-4 After Removing Unwanted Columns and Label Encoding Categorical String Data*

	Survived	Pclass	Sex	Age	SibSp	Parch	Fare	Embarked
0	0	3	1	22.0	1	0	7.2500	2
1	1	1	0	38.0	1	0	71.2833	0
2	1	3	0	26.0	0	0	7.9250	2
3	1	1	0	35.0	1	0	53.1000	2
4	0	3	1	35.0	0	0	8.0500	2

We'll use impyute's MICE algorithm to get the missing values in the Age variable:

```
imputed_data = impy.mice(data)
```

And that imputes the data. Moving on to outlier detection, we'll use z-score to calculate how much each data point deviates:

```
z = np.abs(stats.zscore(data))
```

This gives us a z-score for all 889 data points since we have dropped two rows. We'll drop any row with a z-score greater than 3.5:

```
threshold = 3.5
outlier_rows = set(np.where(z > 3.5)[0])  #getting the outlier
                                           rows
outlier_free_data = imputed_data.copy()
for outlier in outlier_rows:  #dropping the outlier data points
      outlier_free_data.drop(outlier, axis=0, inplace=True)
```

We are almost there. You can try experimenting with different algorithms for each step. We'll now divide the data to X and y, X being the features and y being the target column (Survived):

```
X, y = outlier_free_data.drop(0, axis=1), outlier_free_data[0]
```

Voila! We now have the clean data and separated columns to work with. All we have to do is apply machine learning algorithms to make a predictive model.

Applying Machine Learning Algorithms

As discussed in Chapter 1, we can classify ML problems into three major categories: supervised, unsupervised, and reinforcement learning. These problems can be further classified and dealt with by using the right predictive algorithms.

After data cleaning and preprocessing, further tasks are relatively simple. In this section, we'll explore the use of predictive algorithms by applying them on a classification problem.

But, before applying machine learning algorithms to our data, we need to do one more thing: split the training data and testing data. You don't want to train your algorithm on a dataset and test it on the same dataset; it might give good results on the data you are training on but not the data you are testing on in real life (i.e., it might overfit).

Let's start with code:

```
from sklearn.model_selection import train_test_split
from sklearn.svm import SVC
```

Train Test Split will help you to get some data for the training and testing some data using the untouched test set. I am keeping the test size at 30% of the total dataset, leaving 70% for the training data size. And for now I am using the support vector machine algorithm to create the machine learning model.

```
X_train, X_test, y_train, y_test = train_test_split(X, y,
test_size = 0.3)
model = SVC()
model.fit(X_train,y_train)
```

And now we have a trained model:

```
>>> print(model.score(X_train, y_train))
0.9274

>>> print(model.score(X_test, y_test))
0.7490
```

There is still overfitting, and the test score is much lower than the training score, but we can always optimize hyperparameters to improve this score.

Model Evaluation Methods

We have already explored how to make machine learning models and use different algorithms. Now let's see how to evaluate these models to determine if they are good enough.

The first thing that comes to mind when you want to evaluate a model is to check its accuracy, which can be done using the following method:

Accuracy = total_number_of_true_predictions/ total_data_points

But this method for calculating the accuracy of the model can be deceiving at times. Suppose you have a binary classification problem, where a person with cancer is labeled as 1 and a person without cancer is labeled as 0. In all of your data points, there's one person suffering from cancer. You have trained a model, and now you are testing it for 100 patients. Your model predicts for all 100 patients and says none of them has cancer, (i.e., all of them are labeled as 0). In this case, 99 out of 100 predictions are right, meaning that if you calculate the accuracy using the preceding method, it'll be 99%. However, in this case, we need to be able to identify that one particular case where a person has cancer. And hence we should define accuracy in different terms as per the problem statement.

When evaluating model accuracy, you first need to understand the concepts of true positive, true negative, false positive, and false negative. Our previous example is useful for explaining these concepts. In Table A1-6, the leftmost column is the name of the person being evaluated for cancer, the middle column is the person's actual results, and right column shows the results predicted from a machine learning model.

Table A1-6. *A table of actual results and predicted results*

Person	Actual	Predicted
A	1	1
B	0	0
C	1	0
D	0	1

Table A1-6 shows the following results:

- Person A was actually positive for cancer and was predicted by the model to have cancer, resulting in a *true positive.*
- Person B was negative for cancer and the model predicted that correctly, resulting in a *true negative.*
- Person C has cancer but the model predicted a negative, 0, which is a *false negative.*
- Person D doesn't have cancer but the model predicted that person D has, hence a *false positive.*

Now that you know what these terms mean, in certain cases we prioritize fewer false positives and in other cases we prioritize fewer false negatives. There are some scores that help us to do so:

- Precision = TruePositives / (TruePositives + FalsePositives)
- Recall (TPR) = TruePositives / (TruePositives + FalseNegatives)

  Recall, also known as true positive rate, decreases as the false negatives increase. In the case of cancer diagnosis, it is important to eliminate false negatives because otherwise the person would go undiagnosed.

Confusion matrix: A table to visualize the number of false negatives (FN), false positives (FP), true negatives (TN), and true positives (TP)

Positive	TP	FP
Negative	FN	TN

So, even if you go for multiclass classification, you can diagonally look at the true predictions.

Based on false positives and false negatives, we have other formulas which try to reduce both of them:

- Fβ-Score

  Fβ-Score = (1+β.β)*((precision*recall)/(β.precision+recall))

  Fβ-Score keeps a balance between precision and recall, and β can be changed to give priority to either precision or recall:

  β = 1 for a balance between precision and recall

  β = 0.5 for precision-oriented scores

  β = 2 for recall-oriented scores

- Roc Curve(Receiver Operating Characteristic Curve)

  One of the most effective measures to evaluate a classification model, it tells us the chances of our model predicting the right class. The curve is plotted between Recall/TPR (true positive rate) and FPR (false positive rate), where

  FPR = False Positive / (True Negative - False Positive)

  TPR is the y axis and FPR is the x axis (see Figure A1-4). The maximum possible value of FPR and TPR is 1 and the minimum possible value of FPR and TPR is 0.

  The area under the curve (AUC) is the chance that the model will predict the right value.

The following are the four ROC distributions in Figure A1-4:

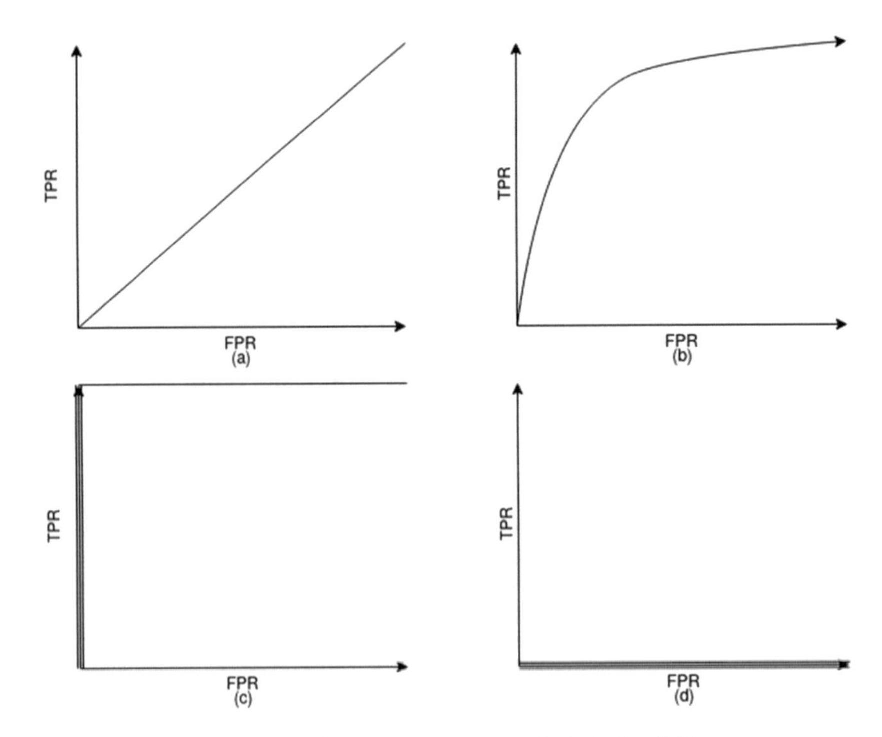

Figure A1-4. *Four plots between TPR and FPR in different scenarios*

(a) Denotes AUC 0.5, which means there is a 50% chance that your model will predict the right value. Hence, the model is of no use and you can throw it away.

(b) Gives you an area of 0.7, which indicates the model has a 70% chance of predicting the right value. Better.

(c) The best-case scenario, where all the values are predicted correctly and there is no false negative or false positive, making TPR 0 and FPR 1.

(d) Informs you that there's a high possibility that somewhere you did something quite silly while writing the code because all the negatives are predicted as positives and vice versa.

Now this sums up almost all the evaluation techniques for classification. Evaluating a regression model is not difficult. Just calculate the R-squared score:

- R-squared Score = (variance calculated by predictions) / (actual variance)

So we can conclude that to make a good model, you must evaluate your model properly.

There's a lot more that can be done, but this appendix has covered some basic aspects of building a machine learning model.

APPENDIX II

Neural Networks: A Brief Introduction to PyTorch and Keras API

This appendix discusses a very basic form of neural network and how to implement it in PyTorch and Keras, both of which we have throughout the book.

If you want to dig deep into the neural networks, I suggest the free online book *Neural Networks and Deep Learning* by Michael Nielson.[1]

A *neural network* is a connection of neurons, each activating when a certain value hits it. It's like a universal function that can adjust its weights and biases according to the nature of the dataset. The number of input nodes is equal to the number of features, and the number of output nodes is equal to the number of target classes. As explained in the first section of Appendix I, the output is one-hot encoded. In Figure A2-2 you can see the number of input features would be five and it would have three classes on output. I will discuss here a fully connected neural network and show you how to implement it in PyTorch and Keras and train the MNIST dataset.

[1] `http://neuralnetworksanddeeplearning.com/`

T. Agrawal, *Hyperparameter Optimization in Machine Learning*,
https://doi.org/10.1007/978-1-4842-6579-6

The MNIST dataset consists of images containing handwritten digits, each image containing a digit and its label. The task in this dataset is to classify the digit from the image. When dealing with images, each pixel is considered as one feature.

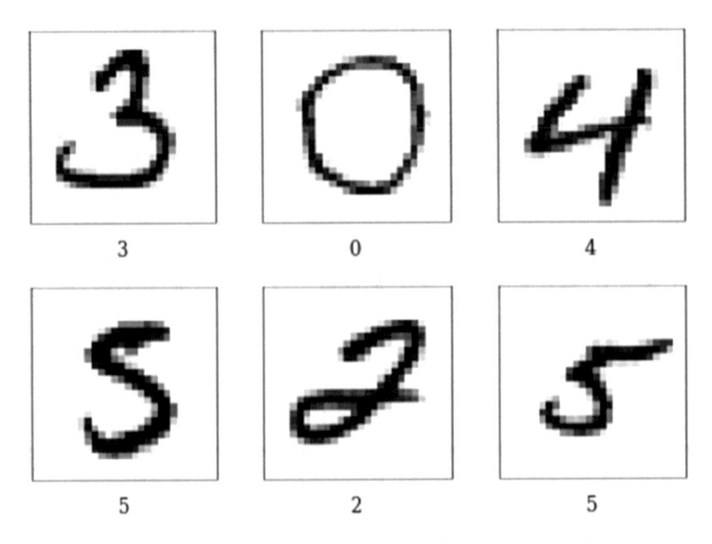

Figure A2-1. *Six random samples from the MNIST dataset*

Before moving further, let's see how neural networks work. I won't get into mathematics, but rather try to give you a conceptual understanding. Initially, random weights and biases are assigned, so if we give the neural network an image of number 3, it'll produce some random output. But as we feed our neural network more and more data, weights and biases (or we can say all the neurons) adjust themselves in order to minimize the loss function and try to perform more specific tasks, cumulatively classifying the number correctly or even performing more complex tasks. We use an algorithm called backpropagation, which changes the weights and biases such that loss is reduced.

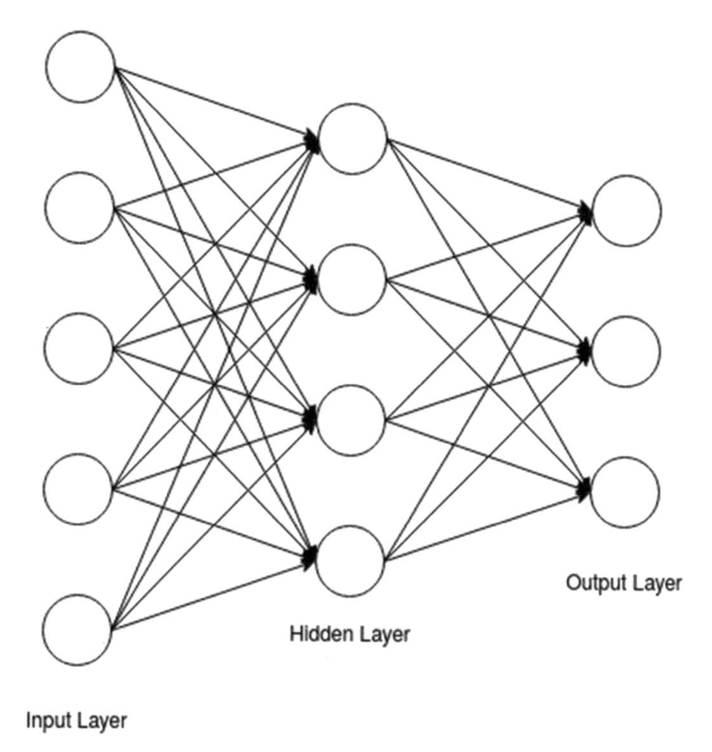

Figure A2-2. *A fully connected neural network with one hidden layer*

Notice in Figure A2-2 how each node in one layer is connected to each node in another—that's how the data flows.

Now let's jump into some coding.

Here we'll implement a fully connected neural network using PyTorch:

```
import torch.nn as nn
import torchvision.datasets as datasets
from torchvision import transforms
from torch.utils import data as D
import torch.optim as optim
import torch
import numpy as np
import torch.nn.functional as F
```

We are using PyTorch, TorchVision, and Numpy. I'll explain all the imports as we go ahead.

```
data_transforms = transforms.Compose([transforms.ToTensor()])
```

```
mnist_trainset = datasets.MNIST('./data', train=True,
download=True, transform=data_transforms)
mnist_testset = datasets.MNIST("./data", train=False,
download=True, transform=data_transforms)
```

First we load the dataset; torchvision provides functionality to automatically download the MNIST dataset using `torchvision.datasets.MNIST`, and we apply the `data_transforms` to convert the data to tensors.

```
batch_size = 1
```

Define the batch_size.

```
mnist_dataloader_train = D.DataLoader(mnist_trainset, batch_
size=batch_size, shuffle=True)
mnist_dataloader_test = D.DataLoader(mnist_testset, batch_
size=batch_size, shuffle=True)
```

Pass training and testing data to `DataLoader()`, which creates a generator object that loads the data in batches, so that we do not have to load all the data at once.

```
mnist_dataloader = {"train": mnist_dataloader_train, "eval":
mnist_dataloader_test}
```

```
dataset_size = {"train": len(mnist_dataloader_train), "eval":
len(mnist_dataloader_test)}
```

We then nicely put them into dictionaries, to use while training and evaluating the model.

```
print(dataset_size)
{'train': 60000, 'eval': 10000}
```

MNIST has 60,000 data points for training and 10,000 for testing.

```
class Network(nn.Module):

def __init__(self):
        super(Network, self).__init__()
        self.net = nn.Sequential(
        nn.Linear(784,512),
        nn.ReLU(inplace=True),
        nn.Linear(512,256),
        nn.ReLU(inplace=True),
        nn.Linear(256,128),
        nn.ReLU(inplace=True),
        nn.Linear(128,10)
        )

def forward(self, x):
      x = self.net(x)
      return F.log_softmax(x, dim=1)
```

Now we finally build a small neural network. We extend the `nn.Module` and in the constructor we define the neural network. There are 784 input nodes since the image size is 28×28, and we flatten it and provide the pixels as features. The output of our network has ten nodes since we have ten classes, 0 to 9; we one-hot encode the labels. Also we have three hidden layers, with 512, 256, and 128 number of nodes, each layer followed by activation function `ReLU()` and output layer followed by Log Softmax Function.

```
net = Network()
```

Initialize the network class.

```
>>> print(net)
Network(
  (net): Sequential(
    (0): Linear(in_features=784, out_features=512, bias=True)
    (1): ReLU(inplace=True)
    (2): Linear(in_features=512, out_features=256, bias=True)
    (3): ReLU(inplace=True)
    (4): Linear(in_features=256, out_features=128, bias=True)
    (5): ReLU(inplace=True)
    (6): Linear(in_features=128, out_features=10, bias=True)
  )
)
```

You can simply print and check the network summary.

```
params = net.parameters()
optimizer = optim.Adadelta(params)
```

We now initialize the optimizer, which will help show weights and biases direction in order to decrease loss.

```
epochs=3
for epoch in range(epochs):
      print("Epoch {}".format(epoch+1))
      for phase in ["train", "eval"]:
            if phase=="train":
                  net.train()
            else:
                  net.eval()
            running_corrects = 0.0
            for data in mnist_dataloader[phase]:
                  net.zero_grad()
                  inp, out = data
```

```
                with torch.set_grad_enabled(phase == 'train'):
                    inp = inp.reshape(batch_size, 784)
                    pred = net(inp)
                    loss = F.nll_loss(pred, out)
                    if phase=="train":
                        loss.backward()
                        optimizer.step()
                pred = np.argmax(pred.detach().numpy(), axis=1)
                running_corrects += np.sum(pred == \
                                    out.data.detach().
                                    numpy())
            epoch_acc = running_corrects / (dataset_size[phase])
            print("{} accuracy: {}".format(phase, epoch_acc))
```

The following are the results:

```
Epoch 1
train accuracy: 0.95015
eval accuracy: 0.9238
Epoch 2
train accuracy: 0.9689333333333333
eval accuracy: 0.9443
Epoch 3
train accuracy: 0.9661666666666666
eval accuracy: 0.9472
```

Now this code may seem a little overwhelming, but all I did was run a loop for epochs, and inside another loop over the data which we got from dataloader object. I reshaped the input size from 28×28 to 784. Passing the flattened image to the neural network and getting it's prediction. After that, I calculated the loss and ran backpropagation. Now to calculate the accuracy, we have the correct observations and predicted values. After finding the correct prediction, we get the accuracy by dividing it by total data size. And you can see the train and test accuracy in each epoch.

Now we'll work on the same MNIST dataset using Keras.

```
import keras
from keras.datasets import mnist
from keras.models import Sequential
from keras.layers import Dense
```

Again, Keras enables us to download the MNIST dataset from its own function. We'll use Sequential to create the network, and Dense layers are nothing but fully connected layers.

```
batch_size = 1
num_classes = 10
epochs = 3
```

We define the batch_size, num_classes, and epochs just as we did while using PyTorch.

```
(x_train, y_train), (x_test, y_test) = mnist.load_data()
```

We load the MNIST dataset.

```
x_train = x_train.reshape(x_train.shape[0], 784)
x_test = x_test.reshape(x_test.shape[0], 784)
```

Again, we flatten the images to 784.

```
x_train = x_train.astype('float32')
x_test = x_test.astype('float32')
x_train = x_train/255
x_test = x_test/255
```

When we loaded data in PyTorch, it was already scaled from 0 to 1, but here the pixel values range from 0 to 255, so we scale them.

```
y_train = keras.utils.to_categorical(y_train, num_classes)
y_test = keras.utils.to_categorical(y_test, num_classes)
```

We convert the labels to one-hot encoding.

```
>>> model = Sequential()
>>> model.add(Dense(512, input_dim=784, activation='relu'))
>>> model.add(Dense(256, activation='relu'))
>>> model.add(Dense(128, activation='relu'))
>>> model.add(Dense(num_classes, activation='softmax'))
>>> print(model.summary())
Layer (type)                 Output Shape              Param #
=================================================================
dense_2 (Dense)              (None, 512)               401920
_________________________________________________________________
dense_3 (Dense)              (None, 256)               131328
_________________________________________________________________
dense_4 (Dense)              (None, 128)               32896
_________________________________________________________________
dense_5 (Dense)              (None, 10)                1290
=================================================================
Total params: 567,434
Trainable params: 567,434
Non-trainable params: 0
_________________________________________________________________
None
Here we define the data; I have chosen the same network
architecture as I did in PyTorch.
```

The first time I call `model.add()` it is acting as two layers, the input layer with 784 nodes and the first hidden layer with 512 nodes. Then the second hidden layer with 256 nodes, third with 128 nodes and finally the output layer with nodes equal to the number of classes. The summary has None written in the Output Shape column, which signifies the batch size.

```
>>> model.compile(loss=keras.losses.categorical_crossentropy,
    optimizer=keras.optimizers.Adadelta(),
    metrics=['accuracy'])
```

model.compile() helps us defining the loss function, which optimizer to use, and what score metrics to use to evaluate the model.

```
>>> model.fit(x_train, y_train,
    batch_size=batch_size,
    epochs=epochs,
    verbose=1,
    validation_data=(x_test, y_test))
```

```
Train on 60000 samples, validate on 10000 samples
Epoch 1/3
60000/60000 [==============================] - 837s 14ms/step -
loss: 0.2971 - acc: 0.9374 - val_loss: 0.2558 - val_acc: 0.9527
Epoch 2/3
60000/60000 [==============================] - 894s 15ms/step -
loss: 0.2476 - acc: 0.9636 - val_loss: 0.2576 - val_acc: 0.9650
Epoch 3/3
60000/60000 [==============================] - 875s 15ms/step -
loss: 0.2163 - acc: 0.9708 - val_loss: 0.2335 - val_acc: 0.9722
```

model.fit() will start the training. We give it x_train and y_train to train upon and x_test and y_test as a validation set. As you can see, in the third epoch the accuracy reaches 97% without overfitting.

You might find that Keras implementation is far easier than PyTorch implementation, but personally I prefer PyTorch, because it's really flexible and easy to experiment with. When you move toward more complex networks like a generative adversarial network (GAN), it's really easy to work with PyTorch and tweak anything you want.

Index

T. Agrawal, *Hyperparameter Optimization in Machine Learning*,
https://doi.org/10.1007/978-1-4842-6579-6

T

U, V

W, X, Y, Z